THE DESIGNS OF BIOLOGICAL FORMS, DEVELOPMENT, AND INITIATION OF CANCER

MIRABOTALIB KAZEMIE

authorHOUSE®

AuthorHouse™
1663 Liberty Drive
Bloomington, IN 47403
www.authorhouse.com
Phone: 1-800-839-8640

First published by AuthorHouse 8/5/2009

ISBN: 978-1-4490-1052-2 (e)
ISBN: 978-1-4490-1050-8 (sc)
ISBN: 978-1-4490-1051-5 (hc)

Library of Congress Control Number: 2009907921

Printed in the United States of America
Bloomington, Indiana

This book is printed on acid-free paper.

CONTENTS

INTRODUCTION

We have a number of important fundamental questions in biology and medicine that are still waiting for answers. One of those questions is related to the designs of biological organisms. How is it possible that many living organisms despite sharing similar regulative systems during their embryonic development and having similar structural proteins differ from each other morphologically, physiologically and behaviorally? A question for which at the molecular level seemingly no satisfactory answer as yet has been found. Another question concerns the initiation of cancer. Our achievements in the areas of cancer-associated mutations and the processes involved in development of cancer are enormous, but still one crucial question has been left unanswered and that is how those mutations instead of causing a cell to die, make it to proliferate. The other questions that we are eager to find exact answers to them are: Is adaptability an active or a passive process? How exactly new living organisms emerge? How instincts develop? How learning faculties and behavior in metazoan develop? Can we speak of 'protein intelligence'?

To provide answers to these questions, I describe in this book for the first time a comprehensive biological theory that I believe to be able to show that the design, behavior, and functions of a biological organism is determined by a regulative program that is encoded by some invariant proteins. I call this design determining program *'primordial program'*. Every biological form has its own specific primordial program; they acquire this program through a process which I name it *'primordialization'*. Unlike many other proteins, the proteins that are part of the primordial programs cannot tolerate mutations. Alteration of any of the proteins involved in a primordial program leads to its destabilization. This can occur in an egg cell, in a cell in the very early stage of an egg cell's development, and in a somatic cell. While the first two occasions end lethal, destabilization of a primordial program in a somatic cell can turn that cell into a cancer cell.

In the light of the primordialization theory the issues of adaptability and variability of the living organisms will be better understood. They are defined as innate potentials in living organisms that serve not only their survival, but also the integrity of their identities. As the reader will recognize, the theory of primordialization takes also a reasonable approach to help differentiate for the first time between the mechanisms which are involved in development and those which are in charge of biological diversity. Regarding 'protein intelligence', I consider it as the supreme kind of intelligence and the basis of cellular intelligence, which in its turn is paramount for the development of instincts, learning faculties, and behaviour in metazoan.

The theory of primordialization like many new ideas in science and medicine has been conceived by intuition; it would provide new insight into the problems that are outlined above. I hope it becomes a source of inspiration for students who are interested in those issues. To put my theory in form of this book I profited from my broad spectrum of education in science and medicine and my familiarity with the subjects as protein biochemist and a former professor for biochemistry and molecular biology. For better understanding of the theory, the chapters of this book are arranged in the following manner: in the first two chapters a detailed description of the theory is given; in chapters three to six I outlined some of the scientific data that the reader needs to know in order to judge upon the significance of the theory; in the chapters that follow the existing notions about the relevant questions are being reinterpreted within the framework of the primordialization theory. The list of books and papers has been considered representative for the purpose and adequate to the size of the book.

In this place, for the convenience of the reader I consider appropriate to make beforehand a brief remark about some terms used in this book. The use of the term *species* is avoided throughout the book, except in one occasion when the definition of this term itself is the subject of discussion. Instead, I use the word *form* to distinguish biological organisms according to their distinct body designs. Under *urprimordial cells* I refer to the very first living cells appeared on earth, and to any other cells that share the same characteristics with those very first living cells. A *primordial cell* I call a living cell that carries the program for the design of the form to which it develops. This brief description, however, is not a replacement for detailed handling of these terms inside the book.

Primordialization

Living organisms differ from each other by virtue of their body designs. An organism with a distinct body design is a 'form'. The design of a form is determined by its 'primordial program'.

Emergence of biological forms requires primordialization: changing of an urprimordial cell into a primordial cell

For the emergence of any kind of biological form an urprimordial cell, a design, and a design-making program is required. I call this design-making program 'primordial program'. A primordial program has to be integrated into an urprimordial cell and turn it into a 'primordial cell'. The process of changing an urprimordial cell into a primordial cell I name 'primordialization'. A primordial cell can then develop into a form, and the program it contains determines the design of that form, as it has been explained below.

Urprimordial cells are the very first living cells that emerged on earth. They must have been able to grow and reproduce; otherwise, they would have died out soon after they emerged. They would have also ceased to exist, if they were not been able to communicate with the environment surrounding them. In order to accomplish this, those cells must have been equipped with signalling proteins on their surface that could establish such communication, crucial in their search for nutrients. In addition to the qualities mentioned above, cells derived from a dividing urprimordial cell must have possessed the potential to remain, under favourable conditions, attached to each other. Further, the pores on the surface of the derived cells must have been able to join end to end and establish intracellular channels. Those were the two prerequisites for the urprimordial cells to turn into multicellular organisms, later.

Forms' designs are encoded by the invariant proteins during the process of primordialization

What kinds of molecules are encoding primordial programs? The answer to this question is this: Primordial programs are being encoded by some of the invariant proteins that are present in urprimordial cells. I consider the following invariant proteins as the very likely candidates among them to take part in encoding the primordial programs: proteins such as those that determine the polarity of organisms; proteins in charge of segmentation and symmetries in developing embryos; growth factors; growth factors receptors; proteins that are involved in signal transmission; and proteins that are in control of DNA molecules. The number of the invariant proteins that are being used for making primordial programs constitutes only a fraction of the urprimordial cells' proteins, but they can encode theoretically thousands of different primordial programs depending on their number, their location inside or on the surface of the cells, and their manner of interaction with each other. In other words, primordial programs differ from each other by the way those invariant proteins are combined and arranged during the primordialization processes. In this way a primordial cell obtains the information for the ultimate appearance, function, and behaviour of the organism to which it develops. The information on the design of a multicellular organism present in its primordial cell needs to be translated into a comprehensive plan that has to be read after each cell division throughout the entire processes of development and differentiation. In other words, each cell in a developing embryo has to be provided with a set of instructions defining its specific duties. How this takes place? At any level of development and cell differentiation the organism's primordial program (alternatively, I refer to it as design regulative program) adjusts the activities of the subordinate regulative proteins. With this the design regulative program makes sure that the cells' activities remain within the boundary of the organism's general design. While the job of the design regulative program is to code the design of the form and preset its boundary, the functions of the subordinate regulative proteins are to lead the cells and organs to develop within that preset boundary. Passing of information in this manner will continue from one cell generation to the other until the structure of the organism is complete.

Forms cannot be transformed

According to primordialization theory no biological form can be transformed into another. When any of the proteins involved in the primordial program of a cell is altered, genetically or epigenetically, this would have disastrous consequences for that cell or even for the organism as a whole, because those proteins cannot tolerate alterations. In case such an alteration takes place in a primordial cell, or an egg cell as its replicate, that cell would either die or give rise to nonviable deformed structures. When alteration of a protein involved in a primordial program occurs in a somatic cell, that cell would turn into a cancer cell. I will talk about this under the heading 'Initiation of cancer' later. Here, I want to emphasis that I prefer not to use the term *transformation* to describe the turning of a normal cell into a cancer cell, because I apply this term strictly with regard to the designs of organisms. Instead, I call such changes *aberrations*; to this I also include any changes that distort the general features and proper functioning of a biological form. For example, an anencephaly that is caused by a defect in a subordinate regulative system responsible for the formation of head structures is an aberration in vertebrates and invertebrates alike, because none of those animals when born with anencephaly is representing a new form; with or without head they are the same organisms. Very different from this kind of anencephaly is the anencephaly in chelicerates, which is not an aberration, because their primordial cells are so programmed that these animals have no heads, and that the organs which are operating in head section in other arthropods are located in their prosoma section [the bodies of chelicerates are subdivided into two sections, the prosoma (front) and the opisthosoma (back). The prosoma contains segments that bear the chelicerae (spider's fangs), pedipalps and four pairs of walking legs. It carries the main feeding, sensory, and locomotor apparatus].

HOW PRIMORDIALIZATION WORKS TODAY?

*It works as reprimordialization using cells
equivalent to urprimordial cells.*

Sources of cells equivalent to urprimordial cells

As it has been mentioned above, the process of primordialization needs urprimordial cells. However, since free-living original urprimordial cells are unlikely to be available on earth any more, cells equivalent to them are being used in this process. Such cells can be produced when, for example, an organism or parts of it disintegrate into undifferentiated cells through disintegration of its original primordial program. However, these cells only then can take part in primordialization process when the amino acid sequences of the proteins involved in their original primordial program are not changed. Then, the proteins from the original program are reprogrammed into a new primordial program. I call this reprogramming process 'reprimordialization'. In the following I describe some sources of cells equivalent to urprimordial cells together with examples of reprimordialization.

1. Some embryonic cells in placentals can be turned into cells equivalent to urprimordial cells. The external cells of their embryos at 8-16-cell stage (the merula) are such cells. Some of them develop into a trophoblast leading to the formation of placenta. I consider these cells equivalent to urprimordial cells for the following reason. Placenta, I believe, is an organism, not an organ. If it were an organ, then it must be integrated into the anatomy of the organism that continues to grow after the birth. Furthermore, the fact that it nourishes inside the uterus independently from the foetus underlines this notion. Reprimordialization requires that the external merula cells in order to be able to lead to the formation of placenta, first of all, their original primordial program must be erased. This means that they become equivalent to urprimordial cells. These cells are then reprogrammed getting a new primordial program, which determines the design of the placenta. It needs to be mentioned that this process should not be mixed with the ability of embryonic cells in some organisms to reproduce offspring, such as the larvae of Trematoda. In this case no primordialization process is required, because the daughter embryos and the embryo that gives rise to them have the same design and so the same primordial program.

2. Cells equivalent to urprimordial cells can be produced trough partial or total disintegration of some metazoan organism under a process that has been called, incorrectly, metamorphosis. Let me explain this by taking Drosophila as an example. An egg laid by Drosophila fly grows into a Drosophila larva, which is a

form with a design totally different from the fly form. The fly, contrary to the current notion, is not the product of further development of the larva, because it has its own distinct design. The design of the fly seems to be programmed inside a group of cells that in the literature are being referred to as imaginal cells [A fertilized egg begins its development with a series of very rapid nuclear divisions with no cell division. Most of those nuclei migrate from the middle of the egg toward the surface, where they form a syncytium. Later plasma membranes grow inward from the egg surface to enclose each nucleus, converting the syncytium into a blastoderm consisting of some 5000 separate cells. Thereafter, the larva emerges and passes through three stages, or instars, separated by molts in which it sheds its old coat and cuticle and lays down a larger one. When the life of larva nears the end, its tissues break up. Its disintegrated epidermis cells are deposited in different segments of that organism (imaginal discs). The organs of the fly develop from this group of cells]. Interestingly, those cells are undifferentiated. I assume that during the disintegration process the larva cells are being cleared from their original primordial program and reverted to undifferentiated cells. Since they have the original proteins still active for reprogramming, they are equivalent to urprimordial cells. As such they can be reprimordialized. This means that they can acquire through rearrangement of those proteins a new primordial program specific for the design of the fly form of the Drosophila. I predict also the existence of a reprimordialization process for the emergance of the larva form of the Drosophila. This possibly takes place at an early stage of the oogenesis in the fly.

3. The production of urprimordial cells is also very obvious by nauplius larvas of parasitic copepod *Haemocera danae*, when they become parasite in a polychaete worm. Inside the host the life of nauplius larva reaches an end; its structures disintegrate, leaving no sign of a larva behind. What is left after this disintegration is an ovoid mass of small undifferentiated cells. This mass of small undifferentiated cells is equivalent to urprimordial cells. These cells then give rise to a new form, a free-swimming organism inside the worm. Here, a reprogramming of the first primordial program into a new one takes place, i.e. a reprimordialization.

In this place I want to substantiate my notion that larvae, no matter to which taxonomic group they belong, are distinct biological forms. They are active, feed themselves, and with some exceptions they live independently. In addition, larvae and the 'adult' organisms can live under very different environmental conditions. A larva can live parasitic, while the 'adult' organism lives free (e.g., Gordioidea), or the larva can live free, while the 'adult' organism is a parasite (e.g., parasitic copepods). Finally, many larvas instead of turning into the so-called adult organisms, can reproduce independently, either asexually (e.g., Cecidomya), or sexually (Cetenophora). It is noteworthy that Darwin himself founded undisputable that larva and 'adult' organism during metamorphosis represent two independent organisms, when he wrote, "No one probably will dispute that the butterfly is higher than the caterpillar."

4. Cells equivalent to urprimordial cells are produced as part of the process that leads to the emergence of medusas from a hydroid polyp. They are produced through disintegration of parts of the polyp. The polyp usually produces its own kind, i.e. other polyps, through budding. Occasionally, some of those buds behave differently and produce medusas. To do this they have to disintegrate and form gametangia. It is during this process that the polyp's primordial program is being replaced by a primordial program specific for the design of the medusa. Thus, the cells inside the gametangia are equivalent to urprimordial cells, in order for the reprogramming to take place.

5. I consider processes such as the one that turns the unicellular amoeba *Dictyostelium discoideum* into a multicellular slug also reprimordialization. When facing starvation, the amoebae assemble into multicellular mounts. The cells within the aggregates then turn into undifferentiated cells. These cells reorganize themselves and lead to the formation of a multicellular organism, the slug. For this to happen the undifferentiated cells have to be cleared from the primordial program of the amoeba (become equivalent to urprimordial cells) and subsequently be furnished with a new primordial program, one which determines the design of the multicellular slug. This example could possibly reflect the

situation when the original urprimordial cells started to aggregate to give rise to the first multicellular organisms on earth.

Egg cells are replicates of primordial cells

Since in Metazoa the regulative program concerned with the basic design of organisms and the strategies of their subsequent development rest inside their egg cells, these cells must obtain those proteins that are essential for the designs of their respective forms always in the same composition and sequence in order to get the green light for starting the development process. For this reason I consider egg cells as replicates of primordial cells. I find a number of findings in the literature that are in agreement with this notion. In mouse zygote it has been found that the first cleavage occurs 17-20 hr. after fertilization, and no mRNA synthesis occurs during this period. In Drosophila, the proteins that determine the initial polarity and the spatial coordinates of the developing embryo are already present inside their unfertilized eggs [It has been demonstrated that egg cells in which any of those proteins are altered produce embryos that fail to survive for lacking large parts of their bodies such as head, thoracic, and abdominal sections].

Functions in multicellular organisms derive from cell's functions in urprimordial cells

Functions in all multicellular organisms take their origin from the functions that urprimordial cells were already equipped with. What primordialization does is the redesigning of those functions in primordial cells in such a way to fit the needs of the organisms that develop from those primordial cells. I want to explain this notion on the example of slug mentioned above. This organism has light sensitive structures at the anterior end of its body. These light sensitive structures are, according to primordialization theory, the result of reorganization and reprogramming of the light sensitive molecules and structures present in the undifferentiated cells (cells equivalent to urprimordial cells) of the mounts. With regard to development of sight in invertebrates and vertebrates, the principle is the same. The eyes of these organisms differ from each other, because they have been designed differently during their primordialization or reprimordialization from the same light sensitive structures in urprimordial cells and cells equivalent to them, respectively. In this sense, in none of those organisms it can be said that they and their eyes and other functions are evolved from each other. In each case, eyes and other organs are built in a way to best fit the design of each form.

Protein intelligence

Being subordinate to their primordial program, cell proteins function with foresight and in coordinated manner.

Protein molecules have the ability to recognize events and memorize their sequence

Until now, intelligence has been considered as an innate mental faculty reserved for man and to limited degrees for some other living organisms. Primordialization requires that intelligence must be universal, because living cells have to be in possession of a very intelligent system in order to be able to recognize and memorize the design of the forms and act accordingly during development. Living cells, indeed, have such systems. These are based on the intelligence of their protein molecules.

In all phases of development in animals and plants it is thanks to the abilities of their proteins that differentiation and proper tissue and organ formation during development proceed according to their design programs. The intelligence of the surface adhesion proteins, for example, enables the differentiated cells to assemble with their kinds. In experiments where cells from different parts of an early amphibian embryo were artificially dissociated and put into a random mixture, it has been shown that the cells in the mixture were still able to sort out according to their origin. Intelligent cooperation among protein molecules on the surface of a living cell is needed if it has to respond to environmental changes successfully. Indeed, a cell is able to receive different kinds of chemical and physical signals from the environment and convey them from its surface along different pathways to several locations inside its body. Bacteria such as *E. coli*, for example, has been shown to sense chemical attractants over a concentration range of several orders of magnitude and respond to extremely low concentrations of nutrients. This is possible, because proteins that act as receptors on their surface can communicate with each other and make jointly an estimate of the concentration of the surrounding molecules.

Protein molecules can program and reprogram the events inside the cell

Protein intelligence is also behind the functioning of biological clocks. These are self-sustaining clocks that give the rhythm inside the living organisms, whether unicellular or multicellular. One biological clock, which is identifiable in all living organisms, is the one that determines the sequence of the events during cell division. Though the duration of cell division may vary from a few minutes to many hours, depending from the organism, the sequence of the events during this process is, however, timely fixed. The presence of biological clocks in all living cells is evidence for the fact that since the emergence of urprimordial cells protein molecules inside the cells interact with each other in an ordered fashion and that they are conscious of the sequence of the events which occur inside the cells.

Intelligence of protein molecules can very clearly be discerned from their ability to manipulate the DNA molecules. As we know, inside the cells there are numerous proteins that are in charge of DNA synthesis, its storage, repair, modification, and utilization as blueprint for protein synthesis. The fact that a cell is capable of repairing damages to its DNA is evidence of the intelligence and foresight of a dedicated DAN-damage-sensing-, and repairing apparatus. Foresight is also involved in preventing damaged DNA entry into the S and M phases of the cell cycle until the damage is repaired. As it has been mentioned earlier, blocking entry into the S phase prevents the replication of a damaged DNA molecule, which otherwise would lead to defect daughter strands. In addition to the DNA-damage-repairing proteins, cells contain protein molecules that are capable of changing the sequence of a DNA molecule by adding new nucleotide sequences to it, or cutting off sections of variable lengths from it. They can remove parts of DNA sequences known as mobile elements from one location of the molecule and insert it into another location of it. By using these DNA elements, organisms have a mechanism available to change the structural architecture of its DNA blueprint. In the single-celled organisms called hypotrichous ciliates dramatic rearrangement of its DNA had been observed. These organisms in addition to their main nucleus have one or more micronuclei. During sexual reproduction the micronuclei from the two partners fuse and give rise to new micronuclei and a macronucleus. As the macronucleus takes shape, not only are the DNA segments between the coding regions removed, but the coding regions put back into correct order, all in a matter of hours. Furthermore, due to the intelligence of protein molecules, the events inside the cell can be reprogrammed. For example, in experiments where the nucleus of a differentiated cell is transplanted into the enucleated Xenopus egg, the nucleus became reprogrammed to resemble that of the normal egg nucleus [A typical amphibian egg is so large that using a fine glass pipette, one can readily inject into it a nucleus taken from another cell. A complete swimming tadpole could be produced, for example, from an egg whose own nucleus, which has been destroyed by ultraviolet radiation, was replaced by a nucleus from a frog red blood cell].

Intelligence of protein molecules seems to reach well beyond their abilities mentioned above. Proteins were found to be able to give themselves the folding they want. Studies on a number of proteins have shown the existence of intramolecular amino acid sequences that are in charge of intramolecular folding. Furthermore, studies on a protein called *split intein* have shown that this protein is capable of bringing together two pieces of different proteins, knitting them together, and then neatly cutting itself out. Another example of engineering ability of cell protein I want to mention is the RNA interference system or post-transcriptional gene silencing. This is an example of an excellent engineering performance of cell proteins. In this system an enzyme produces small RNA molecules called small interfering

RNAs by cutting long double-stranded RNA into pieces of about 21 nucleotides long. Then one strand of each small RNA pieces is loaded onto a protein called *Argonaute* protein, generating a RNA-protein complex that binds to viral targets by base pairing and cuts it.

DNA IS PASSIVE
AND CHANGEABLE

DNA has always been under the control of cell proteins.

DNA has a passive role in cell affairs

In this chapter I want to underline the fact that contrary to the current view DNA on its own has a passive role with regard to organisms' functions and development. I will also present arguments in favour of the notion that the design of the biological forms cannot be altered by mutations of their DNA. But first, before talking about these issues, I have to briefly review some essential characteristics of this molecule. Living organisms use DNA as blueprint for the primary structure of their proteins. It is made of four different nucleotides, and has a double helix structure. The nucleotides are grouped into three-letter "word" called codons; each codon specifies an amino acid in the primary structure of a protein. Specialized cellular machineries copy the DNA into RNA (which has a similar code) and then translate it into proteins. DNA is also used by living cells as a mean of transfer of information encoded in its nucleotide sequence from one generation to another. For this reason during the process of cell division a cell duplicates its DNA in order to distribute it equally between its daughter cells.

At the time when the first living cells emerged DNA must have been one of their constituent parts from the beginning, designed to store information. However, on its own it could not carry this function if it were not for the presence of special protein molecules that made this happen. Its duplication, too, is fully under the control of cell proteins. And in contrast to protein molecules, DNA cannot communicate with the environment outside the cell. As it is described below, DNA is susceptible to environmental factors such as radiation and mutagenic chemicals. Since this susceptibility of DNA is inevitable, organisms have mechanisms available that can correct the unwanted changes in their DNA, as much as possible, before they can cause harm to them.

DNA plays a passive role during development

Egg cells have the capacity to store some of the proteins involved in their development. This makes them able to proceed up to the first cleavage without involvement of DNA. As it has been mentioned earlier in chapter two, in mouse zygote the first cleavage occurs 17-20 hr. after fertilization, and no mRNA synthesis takes place during this period. This means that the whole process that leads to the first cell division is controlled by the proteins already present inside the zygote. In Drosophila and Xenopus eggs it has been shown that the very rapid cycles of DNA replication at the beginning of their development hinder transcription, indicating that up to moment when transcription starts development depends on proteins that are present in egg before fertilization. The role of DNA as a passive player during development can also be inferred from experiments in which the nucleus of a differentiated cell is transplanted into the enucleated Xenopus egg, whereupon the egg cytoplasm was able to reprogram the implanted nucleus to behave as its own nucleus.

DNA has no role during cell division

DNA obviously plays also no role during the events that lead to cell division. This can be inferred from the following experiment: in *Caenorhabditis elegans* during development a protein called *wingless* protein triggers an early cell to divide symmetrically into two daughter cells, which later give rise to different sets of tissues. This protein does so by acting directly on the cell's cytoskeloton, without getting DNA involved in this process. In addition, I want to mention the fact that if DNA were involved in cell division, the cell would not be able to divide in the first place, because ahead of cell division the DNA replicates are already condensed and packed, thus find themselves in a non-functional state. Another fact that points to the passive role of DNA is the observation during cell cycle that a damaged DNA is prevented from entry into the S phase and M phase until the damage is repaired. Blocking entry into the S phase prevents the replication of the damaged DNA, which otherwise could cause change in the daughter strands synthesized on damaged templates. Equally important is a surveillance system known as the spindle check-point, which ensures that chromosomes during the process of cell division segregate accurately by inhibiting chromosome segregation until chromatin separation is complete. It recognizes the presence of chromosomes whose kinechores (spindle attachment sites) are not properly attached to the spindle.

DNA mutations and organisms' designs

Mutations are alterations in the nucleotide sequences of DNA molecules. They can be caused by environmental factors such a radiation, mutagenic chemicals, etc. Since alterations in the sequences of DNA molecules can change the amino acid sequences of proteins, if they left unchecked they can impair the function of cell proteins. In cases where the mutated proteins belong to somatic cells, they can lead to disruption of normal cell functions; this in turn could cause harm to the whole organism, or eventually lead to its death. In situations of DNA mutations in germ cells the matter is worse. Here it is not one individual that will be affected, but all those members of the offspring that inherit the mutated DNA sequences will be suffering from the consequences, ranging from malfunctioning to morphological aberration and death. The later is certain when the mutations of DNA affect proteins that are part of the organism's primordial program.

Living cells seem to be very aware of the harmful effects of DNA mutations; therefore, since the very beginning of life on earth the urprimordial cells have been equipped with measures to respond effectively to the changes caused by mutations. These measures include checking the inaccuracies in their DNA blueprints, purging them from altered sequences, and putting them back into order. One such measure is the DNA repair mechanism. In cases when the repair system cannot deal with the altered DNA, the organism can then try to inhibit, as a precautionary measure, the division of the cells that are carrying mutated DNA sequences.

The means by which organisms can deal with DNA mutations do not end here. Living organisms have also other ways at their disposal to tackle this problem, such as the RNA silencing mechanism, which silences unwanted DNA elements. It must be added that living organisms have not only the ability to correct the sequences of their DNA molecules that are altered by the environmental factors, but interestingly also the ability to alter it autonomously. One form of DNA manipulation under cell control is DNA rearrangement using mobile DNA elements. This includes nucleotide substitutions, large and small deletions, large and small insertions, and translocations of DNA sequences. Some organisms can even change the sequences of their preformed nucleic acids, a process known as RNA editing, common in chloroplast and mitochondria of fungi and protozoan. In other organisms it has been shown that they can produce new sequences, even at a very fast rate. The best known example for this is the production of DNA blueprints for the variable regions of immunoglobulin molecule in vertebrates as part of their precaution to respond to the enormous number of antigens, most of them had never been encountered by the former generations of the organisms in the past [The rate of mutation in these sequences has been estimated to be 10^{-3} per nucleotide pair per cell generation, which is about a million times greater than the spontaneous mutation rate in other genes]. This is in conformity with my theory, because they produce new antibody proteins to maintain their identities; they want to survive as distinct forms.

The high content of non-functional DNA in multicellular organisms is possibly the result of dumping non-coding DNA sequences

The large piles of non-functional DNA in different living organisms suggest in their turn a passive role for DNA with regard to cell functions. In human with a DNA content of 3.2 picograms per cell only a small fraction of its DNA codes for protein – less than one percent. Taking the DNA content of an human cell as reference, the DNA content per cell of wheat, broad beans, and garden anions (7.0, 14.6, and 16.8 picograms, respectively) ranges from about two to more than five times as that. Tulips have ten times as much DNA per cell as humans. The cells of a lungfish have a DNA content 17 times that of human cells. Sixty percent of these DNA piles seem to be non-functional; half of them are cytologically detectable as heterochromatin, and the other half consists of highly repetitive sequences.

It is important to find out the reasons behind the accumulation of large amounts of non-functional DNA in multicellular organisms. Based on the qualities of DNA described before, the following point can be made. Organisms cannot afford to become incapacitated or die each time when a harmful DNA mutation alters the sequence of any of the proteins that is part of their primordial program. To prevent this from happening, they first try to use their DNA repairing system to correct the damaged DNA sequence as far as possible. When this is not sufficient, the other possibility to achieve this would be through excisions and insertions of nucleotide sequences with the help of mobile elements, on a trial-and-error basis, until the original sequence is restored [Mobile elements are DNA sequences that range in size from several hundreds to many thousands of base pairs. They can be moved from one chromosomal location to another]. At the end, when the organism succeeds in restoring the DNA into its original state, it has to dispose itself of those faulty DNA sequences that have been produced during the correction process. One way of doing this would be through breaking it back into nucleotides; but this may cause for the cell more problems if the enzymes inadvertently degrade its normal DNA sequences. A better way of getting rid of the faulty DNA sequences seems to be their accumulation. I assume that this process, which I refer to as dumping of non-functional DNA sequences, is happening routinely inside the cells of multicellular organisms. In the examples given above, the high content of lungfish DNA could be the result of 400 million years of damping altered DNA sequences in the generative cells of this animal. Damping of non-functional DNA sequences can also be inferred from the observations that point to the existing differences in the content of non-functional DNA among living organisms. In addition, intra-group variations of DNA content have also been shown. The amount of DNA inside insect groups, for example, varies by a factor of hundred. Similar variations have been detected among amphibians. A cell of a bull-frog contains twice as much DNA as that of a toad.

DEVELOPMENT

Similarities of embryos among vertebrates indirectly provide evidence for the existence of primordial programs.

Reasons for the similarities of organs during embryonic development between different organisms according to Darwin's theory of evolution and primordialization theory

From the observation that biological organisms share similar structural features and developmental sequences Darwin inferred that many structures in today's living organisms are representing the transformed patterns of the same structures that originally had been existed in their remote ancestors. Quoting Von Baer he argued that, "the embryos of mammals, birds, lizards, and snakes, are in their earliest states exceedingly like one another, both as a whole and in the mode of development of their parts; the feet of lizards and mammals, the wings and feet of birds, no less than the hand and feet of man, all arise from the same fundamental form." He went on and concluded that the "homological construction of the whole frame in the members of the same class is intelligible, if we admit their descent from a common progenitor, together with their subsequent adaptation to diversified conditions." He contended that on any other view, the similarity of pattern between the hand of a man, the hand of a monkey, the foot of a horse, the flipper of the seal, the wing of a bat, etc., is utterly inexplicable. Based on such assumptions Darwin postulated that the degree of similarities between organs of different animals is reflecting the extent of their relatedness. According to him, the stronger the similarity between two organisms the nearer is their relatedness and the more recent the separation of the two organisms from each other. Furthermore, it has been postulated that the embryonic and larval stages are more or less completely the condition of the progenitor of a whole group of animals in its adult state. In other words, if two or more groups of animals, no matter how much they may differ from each other in structure and habit in their adult conditions, pass through closely similar embryonic stages, they are closely related by being descendants of the same ancestors. Referring to the embryos of mammals, birds, and reptiles, it was assumed that these animals are the modified descendants of some ancient progenitor which was furnished in the adult state with bronchi, a swim bladder, four fin-like limbs, and a long tail, all fitted for an aquatic life. In the following we re-examine the reasons for the similarities of organs in the light of primordialization theory.

From the examples mentioned above let us first examine the case of four fin-like limbs which the land vertebrates' embryos supposedly should have. During early development of land vertebrate' embryos what emerges first in sites of the future limbs are small tongue-shaped buds, four identical of them. These buds show no similarity with fins or other forms of extremities that appear at their position at a later stage of development. With regard to the composition of their cells the two pairs of limb buds also seem to show no difference. They are undifferentiated at first, showing no hint of the subsequent skeletal patterns. Their only differences are their location along the body axis. It is only from their position along the body axis that one pair develops into forelimbs or wings and the other one into hind-limbs.

Next, among the structures that have been considered for very long as the markers of phylogenetic kinship between mammals and the jawless fish are the seven cartilaginous gill arches of the adult jawless fish that form the opening through which water flows to provide oxygen for the capillaries of the gills themselves. These structures are also present in developing mammals' embryos, but later are transformed into different structures in adults. The first arch develops into the two bones of the inner ear, the hammer and anvil; the second one develops into stirrup. The rest of them develop into various cartilaginous structures, like thyroid cartilage and trachea cartilage. The idea of evolution of cartilaginous arches in mammals' embryos from the same structures of jawless fish cannot be true. If it were the case that during the past hundreds of million years those structures have undergone such changes that they ultimately turned into different organs, then one could expect that the original structures should have been disappeared from all phases of development in all vertebrates, not that the same structures still linger in their original form in these animals.

Another organ that has been considered as a phylogenetic marker, backed by evolution theory, is the notochord; it builds the longitudinal axis of the invertebrates called cephalochordates. [Noticeably a similar organ is present during the embryonic development of vertebrates]. According to fossil records these invertebrates have been living on earth much longer than vertebrates; the members of their living representative called Amphioxus live still unchanged in many waters today. The finding that their notochord resembles a similar structure in vertebrates' embryo led some evolutionary biologists to believe that these invertebrates are the ancestors of all vertebrates. This belief is unfounded. According to the theory of primordialization the reason for the presence of notochord in both Amphioxus and vertebrates is explainable: Formation of notochord is integrated in the primordial programs of cephalochordates and vertebrates as well, with one difference that in cephalochordates it is supposed to stay beyond their embryonic period of development and in vertebrates only during that period. They all need a notochord structure for the following reason. Organs do not develop all at once. Their proper sequence and positioning within the developing embryo require a body axis for orientation. In developing embryos in vertebrates, notochord precedes the formation of spinal cord, thus providing the embryo with the essential longitudinal body axis. It is involved spatiotemporally in the formation of many other organs as diverse as brain, heart, kidney, segments of muscles, etc. Without a notochord a vertebrate embryo may not be able to develop its nervous system, vertebrae and many other organs, properly.

Furthermore, the same argument that I brought in connection with cartilaginous gill arches is also valid with regard to notochord; if spinal cord in vertebrates were evolved from the notochord of cephalochordates, then the notochord should have been replaced completely in vertebrates. Next, I want to reinterpret two other observations reported by Darwin. The first one concerns the ability of the alimentary canals in the larva to a dragon-fly and in the fish Cobites to respire, digest and excrete. The second one is with regard to a change of function in the Hydra, when the animal turned inside out, so that the exterior surface digest and the stomach respire. The abilities of the alimentary canals in the larva to a dragon-fly and in the fish Cobites to respire, digest and excrete were considered by him as indications for evolution of one organ into another. On the contrary, according to primordialization theory embryonic cells during the development of an organism undergo a pattern of differentiation that fits the design of that form. The organ-specific and other local subordinate regulative systems are bound to strictly follow and carry out the instructions of the design regulative system (primordial program). When it is required that some cells be equipped so to carry three different functions of respiration, digestion and excretion at the same time, the subordinate regulative systems have to carry out this plan during differentiation accordingly. The urprimordial cells were pluripotent, and it is not unusual that certain cells in this organism are differentiated so to cover three instead of one of the functions present in an urprimordial cell. Concerning the change of function in Hydra, it is possible that according to the design of that organism their cells are programmed so to be able to shift their functions.

I want to elaborate further on this issue focussing on the mutations that displace body structures in Drosophila fly. In this insect the decision as to whether an antenna or leg should develop in its head region depends from the presence or absence, respectively, of a protein known as *Antennopedia* protein. [This protein belongs to the group of homeotic selector proteins. These proteins maintain the distinction between body segments; their mutations change structures in one body region into structures appropriate to other body regions]. The observation that in the absence of this protein leg sprouts in the head instead of antenna has been taken as evidence that insects are evolved from multipeds. I assume that this not correct. If antennae were evolved from legs located in the same positions, then the subordinate regulative system for making legs in these locations in multipeds should not continue to exist in this fly, it should rather been replaced completely by the subordinate regulative system that makes antennae. According to the primordialization theory the reason for this phenomenon is at hand. Apparently, the cells at the antennae regions of the head of this fly have the potential to develop in both directions. Normally, the design regulative program of Drosophila instructs the subordinate regulative system in charge of building antenna to build and put antennae in the specified spots of the head. However, in the absence of the latter system leg regulative system put automatically legs in those spots of the head. In other words, a loss of function of proteins responsible for making antenna triggers automatically proteins in charge of making legs to develop legs, instead. The same is true regarding the aberrations that happens due to loss of function of some homeotic proteins. For instance, in segments belonging to the thorax region abdominal organs develop, and reversely in segments belonging to the abdominal region thoracic organs develop. These aberrations do not indicate that thoracic organs are evolved from organs that were abdominal ones in the past, or conversely.

INITIATION OF CANCER

Proto-oncogenes seem to be part of the
organisms' primordial programs.

Destabilization of a cell's primordial program leads to initiation of cancer

Cancer cells share many characteristics of urprimordial cells, such as their potential for indefinite growth, the increased mobility of their cell membrane and membrane proteins, their diminished ability to adhere to a solid surface, and their growing even without attachment to cellular matrix. Nevertheless, I consider them as 'defective urprimordial cells'. How normal metazoans' cells turn into defective urprimordial cells, I describe in the following.

As we know, normal cells, both embryonic and differentiated, become cancerous when one or more of some important proteins, such as growth factors, growth factors' receptors, protein kinases and nuclear proteins are altered. It is for this reason that the normal genes for those proteins are called proto-oncogenes. Normally, one expects that mutations in important cell proteins would lead to cessation of cell functions, but in case of these proteins it is the opposite, their mutations lead to intensification of cell growth. Why they do this? If there were a cancer-making system inside a normal metazoans' cell, the answer to this question would sounded simply like this: those mutated proteins turn that cancer-making system on. However, it is not in the interest of an organism to have such a system of self-destruction. It is also very unlikely to assume that some altered cell proteins are capable of changing the whole chemistry of a cell to the extent to make it cancerous. One way of finding the connection between the proto-oncogens and the intiation of cancer, in my view, is to look into the reasons behind the phenomena of attachment of dividing cells to the extra-cellular matrix, the so-called social behaviour of the cells. I believe that before a cell belonging to a tissue or organ starts to divide, it seeks the approval of the neighbouring cells to ensure that it divides within the context of the general design of the organism, whether this organism is a human being, a monkey, a cat, a mouse, a chicken, etc. It does this using something which looks like an identification card. Once its identification card is being approved by neighbouring cells, that particular cell loosens its contacts again and starts dividing. In this 'identification cards' is the organism's design encrypted. Its 'words and letters' consist, among others, of growth factors, growth factors' receptors, protein kinases and nuclear proteins that are known as proto-oncogenes, the signalling proteins that are stretched from the surface of the cell to the nucleus. Since these proteins are invariant proteins, according to the primordialization theory, they must have part in the primordial programs that determine the designs of the organisms and so the proper place and function of their cells within their specific boundaries. In any cell of a metazoan when a protein that is involved in its primordial program is altered, that cell turns into a defective urprimordial cell. A defective urprimordial cell fails to recognize the primordial program of the neighbouring cells.

Distinction between the process that leads to the formation of Placenta and the process that leads to initiation of cancer

To understand the uniqueness of cancer cell, a distinction must be made between urprimordial cells that are reprogrammable through reprimordialization and those which are not. For the purpose of reprimordialization the proteins that are taking part in the process have to stay intact. This is how the emergence of human placenta (I consider placenta as an organism) from human embryonic cells is possible, as it is explained under the rubric 'Sources of urprimordial cells'. A cancer cell, on the other hand, cannot be reprimordialized, despite being equivalent to an urprimordial cell. The reason for this is because of the alteration of one or more proteins of its primordial program. It is for this reason that I call a cancer cell equivalent to a defective urprimordial cell. Since the surrounding tissue cannot direct and control the growth of a cancer cell within the normal boundaries of the organism's structures, it grows to a cell mass that eventually invades the surrounding tissues and organs.

Adaptability Ensures the Survival of Living Organisms as Distinct Forms

Adaptability is a biological potential that allows living organisms to undergo functional, behavioural, and morphological changes within the boundaries of their forms. Organisms utilize this potential to encounter the impact of the environmental changes, and to survive as distinct biological forms.

Adaptability helps living organisms to survive, and preserve their identities

The potential for adaptation is common among all living organisms. This ability is usually recognizable in the flexibilities and variations the living organisms show with regard to their behavior, anatomy, and physiological functions in response to changing environmental conditions. At the molecular level, adaptability means autonomous rearrangement and reorganization of the organisms' macromolecules in response to the environmental factors that affect them. This involves, for example, modification and repair of DNA molecules, sequence and conformational changes of protein molecules, redesigning of protein molecules, reshuffling of cytoskeleton elements, and readjustment of metabolism.

45

As it has been mentioned before, one form of DNA modification under cell control are DNA rearrangements using mobile DNA elements. This includes single nucleotide substitutions, insertions or deletions of large and small nucleotide sequences, and translocations of DNA sequences. In fact, living organisms have not only the ability to alter their DNA sequences, but also to produce new ones, even at a very fast rate, as it has been described in chapter four in connection with the production of new DNA blueprints for the variable regions of immunoglobulin molecules in vertebrates.

It is important to keep in mind that the changes the organisms introduce to their DNA sequence do not pertain to the sequences of their invariant proteins that are involved in encoding their designs. Therefore, adaptability must be understood as a mean that serves not only the survival of the living organisms, but their survival as distinct forms. I define adaptability as a potential in living organism that enables them to use a variety of measures to survive and at the same time maintain their identities as distinct forms in an ever-changing environment. In each organism the potential for adaptation is part of its primordial program. If the environmental conditions become so harsh that the organism fails to mobilize those measures, then it will die. While according to evolution theory of Darwin adaptation would ultimately lead to evolution of one form into another form, at the cost of identity of the primary form, under the primordialization theory, however, transformation of one form into another is not possible; there are no drives in biological forms to move passively toward their own extinction.

Adaptability is not dictated by the environment

My notion of adaptability differs from the concept of adaptability of Darwin. He supposed that it works as changes that occur at random under the control of the environment, without direction. His point of view led him to the conclusion that, as he put it, "if they [those changes] were allowed to accumulate for many generations they lead to evolution of new forms." According to my primordialization theory, changes brought about by adaptability are not environmentally caused changes. On the contrary, these are the ways how organisms manage to absorb environmental effects. They are manifestation of the organisms' own capabilities, whether it is a change in color or variations in size, shape, etc. Darwin's concept of adaptability is based on his interpretation of the works of breeders. He thought that "they unintentionally expose the organic beings to new and changing conditions of life, by which variability ensues, and among those variations they select those they want through breeding." I do not agree with his notion, because when we look at all dog breeds (one of the examples of domesticated products Darwin was referring to), they all are still dogs. The differences they show in shape or size are being brought about by the existing potentials within that biological form that is dog. To give another example, no matter how many varieties of wheat have been produced by breeders since the dawn of human civilization, they all are still wheat. When a breeder manages to produce, for example, a wheat variety that simultaneously encompasses three different qualities such as high content of protein, resistance to disease, and a relatively shorter maturating time, it is because of crossing between wheat varieties that maintain the potential for one or the other of those traits active at the time of breeding.

In nature, a famous example of adaptation is the case of the salt-and-pepper and black varieties of *Biston betularia* moth. In pre-industrialized England this moth had a mottled-gray upper wing speckled with areas of black. The color and design of their wings provided an effective camouflage against the lichen-covered trees of their living environment. For the first time in 1840 moths with black variation of their wings were observed there. Half a century later, 98 percent of the moth population in the same area were with black wings. According to primordialization theory nothing unexpected happened to those moths, because their primordial program enables them to react in that way. They obtained the black pigmentation because they had the potential for black pigmentation, i.e. its biochemistry, even in the originally white areas of their wings. Under conditions where the industrial pollution has increasingly darkening the bark of trees, on which these moths rest, the use of this potential allows them to adapt to this new environmental change, in order to reduce the chances of being seen by their predators. The same argument can be applied to answer the question why, for example, leaf-eating insects are green, or why animals change their coloration in response to the seasonal or instant changes in the intensity of light impact on the surface of their bodies.

Sexual reproduction

*The potential for sexual reproduction
was rested in urprimordial cells.*

The design of sexual organs cannot be achieved by random mutations

About the origin of sexual reproduction in living organisms there are speculations that it was evolved in organisms that were originally reproducing asexually. On the contrary, I assume that the potential for sexual reproduction rested in urprimordial cells, and for each form the method of reproduction is designed during its primordialization. I want to explain this on the formation of sexual organs in *Caenorhabditis elegans*. It is a simple organism and reproduce sexually.

An adult *C. elegans* is about 1mm long and consists of only some 1000 somatic cells and 1000-2000 germ cells. This animal is transparent and its cells can be watched as they divide, migrate, and differentiate. The cells of the germ line separate from the somatic lines at an early stage of development. It has two sexes: a hermaphrodite and a male. The hermaphrodite sex can be viewed as a female that produces a limited number of sperm. She can reproduce either by self-fertilization using her own sperms, or by mating with a male. The egg-laying orifice, the vulva, is a ventral opening in the hypodermis (skin) formed by cells that arise from this layer. A single non-dividing cell in the gonad, called anchor cell, attaches or 'anchors' he developing vulva to the overlying gonad to create a passageway through which the eggs can pass to the outside world. Laser destruction studies show that the anchor cell is responsible for inducing the three nearest hypodermal cells to the gonad to form a vulva. The inducing signal from the anchor cell ensures that the vulva develops in exactly the right place in re-lation to the gonad. If the anchor cell is killed, these cells, instead of following a vulval lineage, give rise to ordinary hypodermal cells. If all of the gonadal cells except the anchor cell are killed, the vulva still develops normally, indicating that only the anchor cell is necessary for this induction. This information when viewed in the context of primordialization theory means that the func-tion and positioning of anchor cell between the gonad and the hypodermal cells is only possible when it is already programmed in the framework of the general design of the form. For the sup-port of this notion I give the following analyses. Supposedly, if it were a mutation that turned one of the cells into an anchor cell, then, it must have been followed by a second mutation that en-abled the anchor cell to produce a new protein intended to create a passage (vulva) in the hypodermis layer. In order for this struc-ture to be formed, some other mutations within the participating hypodermis cells must have been occurred, too, making them, among others, capable of producing membrane receptors to bind the anchor cell's protein. This is still the beginning. Certainly, many more random mutations and natural selection work would have been needed to make the hypodermal cells understand the purpose of those changes and initiate the necessary intra-, and intercellular responses toward formation of the vulva. This is very unlikely that such a coincidence ever happened. Now, let us assume, for the sake of clarification that at some stage during the development one of the embryonic cells was mutated into a germ-line cell to produce gametes, but there was no anchor cell and no orifice. In a situation like this the animal was actually wasting its energy resources producing gametes that could not get into the outside world. This would not be a sign of fitness for an organism to waste its energy by producing gametes in such an ineffective way. The only way that all morphological and functional aspects of sexual reproduction can be expressed in a coordinated manner is through primordialization of urprimordial cells or cells equivalent to them with a primordial programs specific for *C. elegans,* utilizing the potential for sexual reproduction already present in urprimordial cells.

Sex differentiation

The next point to be discussed is the mechanism behind differentiation of a form into male and female sexes. In any living organism, both male and female sexes develop according to the same form program that is present in their primordial cells. Therefore, the difference between the two must be due to modification of some subordinate programs. In this regard interesting is the finding in *Schizasaccharomyces pompe*, where its cells in addition to transitions between asexual and sexual behavior also show a transition between the opposite sexual statuses. In this organism a shift between two sexually opposite cells has been found to be controlled by the cell type-specific proteins. A cell in order to change its type alters the sequence of its type-specific protein through conversion of its sequence. In mammals the protein which is required for determination of sex is a protein named *SRY*. If this protein is available during embryonic development, it acts on the developing gonads and leads to their differentiation into testes rather than ovaries. It triggers the differentiation of sertoli cells from one of the somatic cell lineages in the indifferent gonads (genital ridge). Once sertoli cells begin to differentiate they must then signal to the other cell lineages that they should follow the male pathway, leading to the formation of functional testes. In the absence of this protein the cells of genital ridge differentiate into folicle (granulosa) cells.

The purpose of sexual reproduction

About the general purpose of sexual reproduction, no one knows exactly. Nonetheless, there is suggestion about its involvement in reducing the effects of DNA mutations. Egg cells may achieve this objective through recombination of its own DNA with a similar set of DNA blueprint coming from sperm. However, it appears that living organisms not always utilize this path. Egg cells can, for example, develop without merging with sperm, a method referred to as parthenogenesis. In rotifiers, nematodes, mites, and insects haplodiploidy seems to determine the sexes, in which haploid males develop from unfertilized eggs and diploid females develop from fertilized eggs, although in the parasitic wasp *Nasonia vitripennis* haploid females have been found, too. Egg development can also be induced by factors such as temperature and mechanical stimuli. In marine worm, Bonellia, the future of an egg whether to become a male or female depends from whether it falls on the proboscis of female or not. If it does, then it develops into a male worm, otherwise, into a female one. In some alligators it has been shown that if the temperature kept experimentally at 30°C, eggs develop into females; if kept above 34°C they develop into males. There are lizards which consist only of female individuals and reproduce without mating. To this comes that many organisms propagate asexually. Large number of plants propagate vegetatively by forming multicellular offshoots that later detach from the parent plants. Likewise, in animal kingdom a solitary multicellular Hydra can produce offspring by budding; sea anemones and marine worms can split into two half-organisms, each of them then regenerate the missing half. It seems that despite some advantages the sexual reproduction may have for the organism, no specific purpose has been found for it.

BEHAVIOUR

Instincts and learned behaviour go back to
urprimordial cells' behaviour.

Relation between survival strategies of unicellular organisms and behavior in metazoan

It is important for their survival that the living organisms interact with the environment in which they live effectively. They must be able not only to distinguish between desirable and undesirable factors, but also to turn away from undesirable factors and move toward desirable ones. Since their metabolism depends from the supply of nutrients, which in their turn are subject to extreme variations, they have to recognize every possible variation in their surroundings and know how to deal with them. In case of single-celled organisms all means and strategies of survival are in one cell, as it was the situation with urprimordial cell. In metazoan the means of interaction with the environment, which were originally present within urprimordial cells, have been reprogrammed and distributed over groups of cells and organs. The primordial program of each organism redirects the urprimordial cells' functions in such a way to cover its own specific needs. The reasons for these needs are obvious. In vertebrates and invertebrates due to differentiation of their tissues most of their body parts have no direct access to the outside world. Therefore, information coming from the environment must be taken by cells, tissues, and organs that are exposed to it and relayed to the rest of the body. At the same time information must be transmitted in the opposite direction, that is from inside to the outside. To this comes that these organisms must analyze the incoming and outgoing information in the same way as urprimordial cells did. In this sense, nervous systems [in organisms with more complex designs] are restructured versions of the urprimordial cells' communication and decision-making systems. When these faculties obtain their ultimate shape at birth, or even after birth, given that no outside help is involved, we tend to call them instinct or unlearned behavior; when these faculties take shape after birth through learning we call them learned behavior.

Behavior like any other functions serves a purpose; for that reason its components must be put together right from the start. For example, let us consider a bird that has to feed its chicks. In order to be able to carry this job effectively, first of all consistency between its abilities of flying, finding food, building nest, and recognizing the cries of its chicks demanding for food must be established. The bird also needs to know that its chicks must be fed on a regular basis for a certain period of time. All this can become possible through the process of primordialization only. The assumption by Darwin that instincts are acquired through accumulation of tiny changes of behavior by natural selection for the benefit of the organism is not satisfactory, because the coherence in sequence and purpose which we see in instincts can not be established among the random tiny changes of behavior. It seems to me astounding that Darwin despite his observation of wide differences between the instincts of neuter ants and sexually active ants still thought that instincts can be evolved by accumulation of unrelated tiny events along many generations. As he himself described, neuter ants were making the working population in an ant community and were being developed from the eggs laid by fertile ants. Not only are neuter ants sterile, but also morphologically very different from fertile ants, such as the shape of their thorax and the lack of wings. In my assumption the difference between the instincts of neuter and sexually active ants can be explained by reprimordialization: At some level of development in one group of eggs the original primordial program of the fertile ants are reprogrammed into a new primordial program, which then leads to the development of neuter ants. Referring also to the relation between larvae and 'adult' organisms, which I have described in chapter two, I want to emphasis that instincts are not established through accumulation of tiny changes of behavior by natural selection. The larvae have their own specific behavior and 'adult' organisms their own specific ones. At the moment a larva disintegrates, its instincts disappear with it. When through reprimordialization an 'adult' organism emerges, it is furnished with a completely new set of instincts.

Some examples of instinctive behavior showing foresight

An excellent example of instinctive behavior described in the literature is that of the solitary wasp. This wasp before laying an egg, first, build a burrow in the ground; then, in order to secure the right source of energy for her offspring's development catches a caterpillar and place it in the burrow. After laying a single egg on that caterpillar she closes the mouth of the burrow and leaves the egg unattended. The new wasp that emerges from the burrow, without ever knowing her parent or having watched any other wasp, manages to fly and find nutrients. When the time comes for her to lay her own eggs, she proceeds exactly the way her mother did, providing her eggs with sustaining nutrient sources, when possible with caterpillars. She knows about the action of her poisonous sting when she uses it on caterpillars. She knows how to place her egg on a caterpillar. She obviously knows that a caterpillar is a good source of food and will suffice the needs of her egg during its development. She knows how to prepare the proper cement for closing the burrow. This complex and coherent pattern of behavior which foresightedly guarantees the survival of this wasp must have been designed and put together in one step. If one extrapolates this pattern of behavior to human dimension, it would require years of training for an adult human individual to learn abilities equivalent to this. Another example of extinctive behavior in which foresight is involved is the reproductive behavior of treefrogs Chiromantis. They lay their eggs in a tree above water. As the eggs are produced, the mating frogs use their feet to beat the eggs and seminal fluids into froth. In this protective environment the eggs fertilize and develop into larvae. When the foam drops down from tree branches, they carry with them the larvae into the water, where they can reprimordialize into frogs. Another strikingly complex form of instinctive behavior with high degree of foresight is that of caterpillars that develop from eggs of butterflies Nymphalis. First, they glue themselves to a leaf or twig by the cremaster (a spiny process at the end of the body) hanging head downward. If it were not for their foresight, at the time the outer shell of the caterpillar comes off the emerging pupa would fall to the ground. Surprisingly, this does not happen, because the cremaster with which the caterpillar glued itself to the twig or leaf does not come off with the shell, so it keeps the pupa still fixed in the same position. The pupa subsequently reprimordialize into a butterfly, which emerges from the capsule and flies off. This time the capsule comes off together with the cremaster, otherwise the butterfly could not fly away.

Instincts and competence for survival at birth

After getting some insight into the nature of some instinctive behavior, I want to point to an apparent connection between instinct and the level of competence for survival at birth. It seems that the more a living organism is equipped with instincts the more capable it is to survive. For instance, if we compare the number of instincts of a newborn human baby with a newly hatched solitary wasp and look at the number of their skills at that point of their lives, we will find a clear correlation between the two parameters. The only instinctive behavior that newborn human babies show at birth is sucking; so with respect to skills they are very helpless. They are even unable to search for food on their own for years after. On the other hand, as mentioned above, a newly hatched solitary wasp thanks to the high numbers of instincts that it has, posses so many skills that are enough to serve her all her life. A connection between the number of instincts and self-sufficiency in this wasp could not exist, if the instincts were the products of accumulation of tiny changes. Thus, the existing connection between the two must have been programmed at the time of its primordialization.

Humans have been learning the means of survival from animals

Man and those animals that are equipped with only a small number of instincts have to learn the techniques of survival from those animals that already have such techniques; either through direct mimicking, or by means of learning from the experiences of the members of their own populations who learned those techniques before them. On the basis of evidences we have today, numerous patterns of behavior in an adult human being, no matter how complex they look, are already identifiable among many insects and animals. Some of the techniques we are using have been utilized by them for many million years. For example, the beginning of our skills to construct bridges may go back to a few thousand years ago; ants however can build even living bridges perhaps for many million years. Our using of high frequency sound waves to identify objects hidden from our eyes may be a century old, but bats, for example, have been able to produce and use such waves perhaps for millions of years. Some moths upon which bats feed are even in possession of a sonar jamming system. This mechanism helps the moths to escape bats when they close in on them guided by their sonar system.

BIODIVERSITY

Primordialization vesus natural selection

Darwin-Wallace's arguments on biodiversity

According to the Darwin-Wallace theory of evolution the present-day living organisms have been evolved from other living organisms that were living in the past. The reason for this evolution in Darwin's words is that "in every generation some individuals of the same species are better adapted to their environment than the others, and when this continues over long span of time, new species would evolve different from the ancestral species." He assumed that due to the ever-changing environment those organisms that are better adapted have a higher chance to survive and reproduce successfully, thus pass their superior qualities on to their offspring, and as a consequence, each generation will be slightly different from the last. He believed further that variations useful in some way to each being in the great and complex battle of life are preserved while those harmful are destroyed. To explain how useful variations occur and how are they preserved, he assumed two kind of forces, one which causes variations and exists outside the living organisms and the other, natural selection, as he named it, determines which variations are favorable and therefore should be preserved. With the help of natural selection, he believed, it will be possible that a flower and a bee slowly become, either simultaneously, or one after the other, modified and adapted to each other by the continuous preservation of all the individuals that presented slight deviations of structure mutually favorable to each other.

In Darwin's view, environment is acting in two ways: first, directly on the whole organism or on certain parts of it through increased use and disuse of those parts; and second, indirectly by affecting its reproductive system. With regard to disuse of parts he argued that this would lead to their reduced size. He thought, for example, that this happened to the wings of penguins and ostrich. The former because of living on the ocean islands have seldom been forced by beast of pray to take flight, ultimately lost the power of flying; the latter because of the increased size and weight of its body became incapable to fly. Comparing a lizard with a snake, he believed that due to disuse the snake lost its limbs. Concerning the effect of increased use of organs he referred, for instance, to the size and shape of giraffe's neck and tail, and the long beak of hamming birds. The first ones, he argued, occurred as a result of sustained effort for better reaching the canopies for leaves and to drive away the flies, respectively, and the second one resulted from a tendency among the ancestors of this bird to adapt their beaks to the form of the flowers from which they collected nectar. With respect to indirect changes of reproductive system he thought that this kind of variability is inducible because of the extreme sensitivity of the reproductive system to environmental changes. In the following example he tried to explain what he meant under the extreme sensitivity of the reproductive system to environmental changes. Looking for explanation for the size of jaws and teeth of man, he assumed that the free use of arms and hands have led in an indirect manner to modifications of those structures: "The male forefathers of man were furnished with great canine teeth, but as they gradually acquired the habit of using stones, clubs, or other weapons they used their jaws and teeth less and less until their jaws together with their teeth became reduced in size." Darwin's arguments mentioned above cannot hold, because as we know there are no ways that changes acquired by a developed individual through disuse or increased use can affect germ cells; therefore, it is impossible for acquired changes to be transferred to an organism's offspring.

Biodiversity as viewed by primordialization theory

Looking at the morphology and behavior of giraffe, humming birds, penguins, ostrich and man, not through Darwin's theory, but through my theory, we would come to the following conclusions. Giraffe has a long neck, because it is the way its primordial cell is programmed; therefore, it must be understandable that this animal with its long neck favors the leaves at the top of trees than the low vegetations on the ground. Because of the unique design of the animal, the latter alternative would be strenuous for it. Concerning the humming bird, let us assume for a moment that the imaginative ancestors of these birds were originally seed-, or fruit feeders that due to a long-lasting drought period were forced to migrate out of their habitat; then, after exhausting search for food they ultimately found a place where plenty of nectar producing plants were available. What happened next? They would have died if stopped continuing their search, because there is no mechanism available in nature that could have changed the entire anatomy and physiology of those birds to make humming birds out of them. First, in order to reach the inside of the flowers, they had to develop long beaks. Second, they had to maintain a steady position in the air, for this they had to reduce the size of their bodies and at the same time increase the stroke frequency of their wings. The conclusion that I can draw for humming bird is that they have no ancestors looking different from them; they were emerged with a body design perfectly suitable for collecting nectar. The design included all the necessary morphological, physiological and biochemical prerequisites characteristic for a humming bird. Regarding the wings of ostrich and penguins, we have to realize that wings are needed not only to help birds fly above the ground, but also to serve birds like ostrich to get more upwind in order to run faster. Wings can also be designed in such a way to be fit for swimming, as it is the case with penguins. One should bear in mind that it is not only the presence of wings and feathers that characterize these birds, but also the wholeness of their forms including their physiology, morphology and behavior that make them penguin. To get to humans, there is no way by which the jaw and teeth of man could be evolved, directly or indirectly, from those of an ape. It is also unthinkable that man had forefathers that were furnished with great canine teeth, and they gradually lost the original size of their teeth together with the size of their jaws. It is impossible that changing environment could change the anatomy of a living form. Behavior and structure in all metazoan are programmed together through primordialization.

Similarity of sequence between homologous proteins and estimates of the rate of DNA mutations provide no evidence of phylogenetic relationships

Recently protein sequence analyses have been used to construct phylogenetic relations among living organisms. It has been assumed that the stronger the similarity between the sequences of two homologous DNA and/or proteins, the nearer is their relation; conversely, the bigger the difference between the two sequences, the further apart they are from each other. Beside this, in some other studies attempts have been made to figure out the rate of DNA mutations in living organisms. Assuming that DNA mutations occur at a regular rate, it has been expected to find a proportionality between the time elapsed after branching of related organisms from their last common ancestor and the number of mutations fixed in their DNA by natural selection. In the following, I will bring arguments to show that these assumptions cannot be correct.

1. The differences found in the sequences of homologous proteins in living organism do not reflect the differences they show with regard to their morphology, physiology and behavior.

For example, the difference between the amino acid sequences of corresponding proteins from apes and man seems to range, according to the type of protein, from zero to less than one percent, but despite this small difference, they do not look alike or behave alike. Similarly, according to a comparison of protein sequence data of rat and mouse the observed protein changes are not related to the difference between those two biological forms. In another example, on the basis of similarities found between cytochrome c sequences of cow, pig and sheep they should be near relatives; as well as camel and whale, for that matter. Furthermore, as with haemoglobin it has been shown, the function of various mammalian haemoglobins does not vary significantly, despite differences between their sequences.

2. Living fossils - living organisms that inhabit the earth for a couple of hundred million years or more - are still the same as their early ancestors, despite the fact that their proteins must have undergone countless changes for such long periods of time.

3. Regarding the estimates of the rate of amino acid substitutions I want to mention this. When substitutions in the sequences of proteins, as I mentioned above, are not involved in diversification of living organisms, it is evident that estimating their rates has no relevance to this problem. Nonetheless, for the sake of completeness of arguments I discuss this matter below on a number of proteins that have been used for calculating such estimates.

A)   A given amino acid site in a protein may be changed more than once with the result that there will be only one observed amino acid difference where there were two mutational events, if the second change is a return to the original amino acid. Similarly, the organisms that are compared with each other may have incorporated similar changes at the same site resulting in no difference.

B)   Sequence analyses of proteins do not reflect the entireness of mutations occurred in a protein. For instance, in case of haemoglobin it has been shown that changes that occur in the interior of the molecule are generally harmful, and most changes occurring in residues on the exterior of the molecule appear to be less harmful. This is so, because the substitutions that occur in exterior sites of this protein have no impact on its function, while substitutions in interior sites impair its function. Often, mutations that occur in the interior sites of protein molecules can be lethal. Mutations that are lethal as a matter of fact do not pass from one generation to the other, thus cannot be preserved. In cytochrome c about 74 to 81 of its amino acid residues are variable, and their substitutions apparently do not affect the function of the protein, while 29 of its amino acid residues are invariant. The latter are important, because they are needed for combining with the heme group, or for interacting with cytochrome c oxidase, etc. Histone-IV protein apparently shows only two substitutions in peas and cattle, because virtually all amino acids residues in hestone-IV are invariant.

C) The matching property of homologous protein sequences may be thrown abruptly out of phase by deletions of one or more amino acids. For instance, human and carp haemoglobins differ from each other in that site 47 is occupied by an alanine residue in the carp alpha-chain and is a gap in human.

D) Because of the degeneracy of the genetic code, there are 61 amino acid-specifying codons, instead of 20. Codons that specify the same amino acid are called synonymous codons. These codons differ from each other in their third nucleotides. Mutations at the third-position of synonymous codons cause no amino acid substitution in the sequence of a protein.

From the data provided by sequence analyses one can draw these conclusions: 1. the existence of phylogenetic relations cannot be proven. 2. Proteins in living organisms, whether they have undergone sequence variations or not, seem to have one common origin, and that must be urprimordial cells. 3. Variations detected in protein sequences between different organisms do not account for their morphological differences. 4. No single form specific protein has been found; something that according to primordialization theory as a matter of fact do not exist. 5. From the perspective of primordialization theory, very important is the identification of conserved (invariant) proteins. These are the proteins that, as I assume, have among them those proteins that are part of the primordial programs in living organisms, determining their design. As it has been described in chapter one, designing of forms is based upon proteins which are invariant and already present in urprimordial cells.

'Form' versus 'species'

The facts described above underline the need that the classification of living organism has to be based on criteria that distinguish forms from each other. The present choices of criteria for classification of living organisms are subjective. The system of naming and classifying living organisms that is still in use today is created by Carl von Linné. Based on the degree of outer resemblance he grouped the living organisms in different categories or taxa, starting from the more basic category named species, toward exceedingly larger categories: the species into genera, genera into families and these into orders, orders into classes, and the latter into kingdoms. Species was defined as distinguishable groups of genotypes that remain distinct in the face of potential or actual hybridization and gene flow. It has been turned out, however, that hybrid speciation, even without polyploidy, is more common in plants and animals than were assumed. Now, according to the reports in the literature there are about a dozen different definition of the concept of species, pointing to the vagueness of this term.

Not only are the criteria used for assigning organisms into species or subspecies categories are vague, but also those used for placing the organisms into higher taxons. Based on their anatomy nematodes were placed close to the 'root of the tree of the animals'. They lack a body cavity called a coelom, which is found in molluscs, insects and vertebrates. Using molecular data they were put together with insects to a group called Ecdysozoa [its members grow by moulting their outer layers]. To justify this shift, it has been assumed that nematodes apparently lost their coelom.

As we see the criteria by which the taxonomists judge the degree of resemblance between organisms depend from the taxonomists' perception about their importance. If taxonomists find a trivial character to be common among a great number of forms, they use it as one of high value; on the other hand, when an outstanding feature happens to be less common, they use it as one of subordinate value. To this adds the confusions caused by notion of phylogenetic relation among living organisms. By applying this notion it could happen that a trivial trait be considered mistakenly as evidence of relationship between some living organisms. To name a few: the manner in which the wings of insects are folded, the mere difference in color in certain algae, the nature of the dermal covering (hair or feathers) in vertebrates, and tiny middle ear bones in mammals etc.

Using differences between homologous sequences of DNA as an instrument of classification of living organisms is equally inadequate. So far some 1000 different organisms' genomes have been sequenced. Though the data obtained are worth spending time and effort, but lacks the least consistency with regard to establishing relationships. As one can see, neither anatomy nor sequence data establish something which does not exist, namely phylogenetic relationships. Changes in sequences of DNA, in whatever large number it might occur, have not been found to be related with anatomical tributes of the organisms. Further, the existing variability of satellite DNA sequences among living organisms led to the belief that this fraction of DNA might be related to speciation. These variations, however, are non-specific even to the extent that members of the same living organism vary with regard to their satellite DNA. The structural and numerical chromosome changes also cannot explain the differences between living forms. First, they are very irregular and second, if there were any connection between them and speciation, one would expect those connections to be identified. In experiments where the effects of change in the number of chromosomes were studied in salamanders, it was found that animals with haploid and pentaploid set of chromosomes were of the same size and appearance; they also did not differ with regard to the morphology of their internal organs [Sections were taken from pentaploid (55 chromosomes) and haploid (11 chromosomes) salamander. While the first sections had cells that were roughly five times as big as those of the second one; the tissue in the pentaploid animal contained roughly one-fifth as many cells as those in the haploid animal. Therefore, reduced cell numbers are compensated for increased cell size]. This observation is consistent with my theory of primordialization that indicates that neither alterations of the sequences of DNA, nor changes in the number and size of chromosomes can lead to evolution of one biological form into another.

From the above deliberation I conclude that the classification of living organisms must be based on criteria that can distinguish the forms definitely from each other. After this has been established, then different populations of any form should be catagorized according to traits such as size, color, etc.

Fossil finds and the problems associated with them

In this part I undertake a critical review of some major vertebrates' fossil data and their interpretations by palaeontologists, whereby I will try to differentiate between assumptions that are based on facts and those which are pure speculations. Fossils when well preserved and their dating is reliable can give us valuable information about the structures and designs of the living organisms that inhabited the earth in the past. Fossil finds are also important in telling us about the way of life of those organisms and the events that affected their lives. Despite all this, one has to remember that the records of fossil finds are not continuous, and they are characterized by gaps, which are generally very large (ranging between dozens of million of years to over hundred million of years). For this reason, as my arguments concerning the major vertebrate fossil finds described below will show, one should hold back from using fossils to 'reconstruct phylogenetic relations' between living organisms; this is in addition to the reasons that I outlined with regard to DNA and protein sequences.

No evidence of Phylogenetic relation among chordates, fishes, amphibians, and reptiles exists

Vertebrates' fossils constitute a major focus of interest for palaeontologists. It has been thought that the remote ancestors of all vertebrates including man are the chordates, animals with a hollow nerve cord, gill slits, and notochord. The oldest fossils of chordates dated back to 600 million years ago. Those fossils are very similar to the living lancelets known as Amphioxus.

Chordates on their way to vertebrates, as it has been assumed, had given rise to the early craniates like hagfishes. From those animals fifty million years later supposedly the first vertebrates have been evolved, characterized by having segmented cartilages protecting their spinal cords, with lampreys as their living representatives. Paleontologists assume further that those early vertebrates then gave rise to vertebrates with jaws and paired fins. The earliest representatives of the latter dated back to 400 million years ago, with sharks as their living representatives. Those animals in their turn are believed to be the ancestors of the bony fish, the first vertebrates with internal skeleton and lungs or swim bladder. It has been further postulated that subsequent diversification of bony fish led to evolution of ray-finned and lobe-finned fishes. While further diversification of the first ones supposedly led to modern fish varieties, the diversification of lobe-finned fishes led to amphibians, the first tetrapods on earth, 370-360 million years ago. The assumptions reviewed above are, however, very speculative as the following analysis will show.

1. If any of the new structures that replaced already existing structures in any of those ancestors was making them more suitable for survival, why should animals with the older structures still exist. With or without cranium, with or without jaw, with or without inner skeleton, with or without lungs, their representatives are still living in waters all over the world.

2. The speculative nature of these assumptions is also reflected in the inconsistency about the age of the fossils. The oldest fossil of hagfish found dated back to 350 million years ago, while those

of the first vertebrates and jaw fish are 500 and 400 million years old, respectively.

3. How did a bony fish acquire lungs, and why this had to be selected? Some palaeontologists tend to believe that there must have been a violent alteration of seasons leading to scarcity of fresh water, so that the bony fishes found themselves in a stagnant pool and suffering from oxygen shortage, having no other choice but to develop lungs. But sustained drought is impossible to be the cause for lungs to develop. The most likely consequence of a sustained drought would have been extinction of those fishes.

4. With regard to the limbs of the first vertebrates one has to ask this hypothetical question. Did limbs evolve while those fishes were still living in water, or later when they moved on land? If it happened in water, there was no benefit in this for them; if this change took place on land, then why they had to leave the waters in the first place. In addition, limbs cannot evolve because of any excessive use of fins for walking; developmentally it is impossible. According to primordialization theory, however, fins and limbs each belong to a different form, which are designed and emerged independently.

Fifty million years later, supposedly, a small creature named stem reptile evolved from among amphibians to give rise to reptiles known from Mesozoic era (covering 150 million years) and the ones that appeared later. Those were reptiles of different sizes and appearances such as dinosaurs, the various types of marine animals like marine turtles and sea snakes, crocodiles, flying reptiles, etc. Here again, the suggestion of an existing phylogenetic relation between amphibians and those creatures is unfounded for a number of reasons.

1. Common sense does not accept this assumption, because why animals that had both water and land as living environment at their disposal should abandon that double opportunity and restrict themselves to land only and become reptile.

2. The assumption that some of them retreated for the second time to water (to build, supposedly, the marine forms), to the same environment that was not good enough for them at the beginning, adds to the unlikelyness of the idea that those amphibians left water in the first place. Therefore, the assumption of a phylogenetic relation between amphibians and reptiles is a very speculative notion.

In the absence of a common ancestor for reptiles, as a matter of fact, the existence of phylogenetic relations among reptiles themselves is also out of question. Of course, they share a number of characteristics like the structure of their jaws, vertebrae, etc., but the existence of similarities between parts of their anatomy in my understanding is not justified to draw the conclusion that reptiles are phylogenetically related, i.e. the later forms were evolved from the earlier ones.

The assumption about evolution of birds and mammals from reptiles is erroneous

From the fossil finds it has been assumed further that the earliest ancestors of birds and mammals were lizard-like creatures that first appeared in Carboniferous period 310 million years ago. They then supposedly divided into Diapsid (sharing two temporal fenestrae) and Synapsid (sharing the lower temporal fenestra only) reptiles, whereby birds became part of a Diapsid clad, and the mammals part of a Synapsid clad. According to a different guesswork birds and mammals, because of sharing the homeothermic characteristics, were considered as sister-groups diverged from a common ancestor about 225 million years ago. Both assumptions have one thing in common: they are only guesses, as the following arguments will show it.

A bird by no means can evolve from a reptile, no matter what anatomical similarities they might have in common, because the design of a bird requires a totally different program than that of a reptile. In order for any reptile to become a bird, that organism must have acquired all anatomical, physiological and behavioral aspects that are characteristic for a bird, simultaneously. This is something absolutely impossible to have taken place. The major characteristics of birds' skeleton are the hollow bones, three functional toes, half-moon-shaped wrist bone, fused clavicles, keeled sternum and short tail. Feathers and claws seem to be found on fossils of some dinosaurs, dated back to 150-160 million years ago. There are also suggestions that the size of avian genomes, smaller compared to other amniotes, had helped birds to fly by reducing the metabolic costs associated with having large genome and cell size. However, small genome size was also common among some reptiles before emerging of birds. This adds one more to the birds' attributes such as feathers, claws, pulmonary characteristics, and parental care and nesting to be possessed by some reptiles before them. However, the existence of dinosaurs with claws, feathers and other birds' attributes neither made them birds nor ancestors of birds. Thus, the only way for birds to emerge must have been through a process which gave them all those major characteristics as essential components of their design, simultaneously. Such a process is nothing else but primordialization.

We turn our attention now to 'mammals'. First, it needs to be pointed out that mammals is a name given to a large group of diverse animals to indicate their having milk-producing mammary glands. Since mammary glands developmentally turn into functional state years after birth, they should not be used as a criterion of relatedness of all those animals. The other assumption that mammals are related because of the similarities of their detached tiny middle ear bones is also weak. Second, there has been not a single fossil which could show a link between any of the various forms that are included in this group. The living monotremes represented by the duckbilled platypus and the spiny anteater or echidna, are egg-laying animals. The marsupials are a diverse group of animals that have been called so because of their females having a pouch in which they carry their immature foetuses. The placentals consist of organisms that are in many respects different from each other. They are grouped into eighteen orders which include insectivores, edentates, rodents, logomorphs, bats, primates, artiodactyls, perisodactyls, carnivores, elephants, cetaceans, etc. The assumption that all these animals have as their predecessor an insectivore (animals not very much different in size from Tupaia), which apparently was living in abundance during Cretaceous period is unfounded. Not only there is no evidence which could indicate that these orders have any phylogenetic relation with each other, but also no proof that the animals in each of those orders are phylogenetic related to each other. For instance, in all those various forms of animals that are classified under artiodactyls (even-toed hoofed mammals) the resemblance of their limbs simply means that the design of such limbs fits the design of the animals to which they belong, nothing more.

The phylogenetic reconstruction attempts also seem pointless when one looks at the primate fossils. It has been assumed that the organisms included in the order of primates such as monkeys, lemurs, man, apes, etc. are the descendants of an extinct prosimian which was presumably was living during Paleocene in Africa 70 million years ago. Oddly enough, such assumptions are being made despite the fact that in Africa following *Eocene* no fossil record of prosimians was found. Furthermore, there exist no fossil which could link the prosimians themselves to insectivores; at least a gap of twenty million years exists between the recorded fossils of insectivores and those of the early prosimian.

Primordialization theory and the emergence of man

Regarding the emergence of man it has been assumed that man is evolved from ape-like creatures, supposedly in Africa during Pliocene and Pleistocene. However, African apes have no fossil record in Pliocene and Pleistocene. Without any fossil record of apes in Africa the postulate of human evolution from ape-like creatures cannot hold. It is useless to rely on a fictive story like this: At least one African ape left the continent about 18 to 20 million years ago, but then after its descendants eventually gave rise to an array of early apes in Asia and Europe one of those Eurasian apes moved back to Africa by about 10 million years later and became the ancestor to the living African apes. Later due to some hypothetical events some of those apes supposedly came down from the top of trees, whereby in the new environment their struggle for survival required increased use of their front limbs. At the end, this supposedly led to a new creature that could walk on two limbs, referred to as australopithecine. These assumptions are very speculative for the following reasons.

1.   Even if it were possible to prove that some 3-5 million years ago there was a climatic change, or any other event such as a volcanic eruption in Eastern Africa, leading to shrinkage of forests, it is unlikely that apes instead of retreating into the still remaining forests dispersed themselves into the open plains. If this were what the apes did, most likely they could not survive very long.

2.   By comparing a very large number of matching genes from human and chimpanzee it has been indicated that favorable mutations in chimpanzee genes were higher in number than similar mutations in human genes. This kind of finding in the context of primordialization means that those alterations have been taking place under the control of the organisms and not the environment. Otherwise, if human were supposed to be evolved from

chimpanzee, then there should be no favorable mutations for the latter and this organism should stay where it was. What would be the purpose of those favorable mutations, if one assumes that they represent the selected ones by natural selection from among the randomly occurring tiny changes? Is natural selection on the path of planning to transform for the second time chimpanzee into a different organism, better than human? Certainly not! However, without any doubt, those favorable mutations are offering the chimpanzee the chances to become stronger in maintaining its own form identity.

3. The oldest bone remains which some paleonthologists consider as evidence of an upright walking ape is the three-million-years-old bone remains of an ape-like creature known as Lucy. It is the skeleton of a young adult female creature with over three feet tall; her skull is extremely fragmentary and has an ape size. The youngest bones belonging to a similar creature is about one million years old. In all those cases perhaps the shape of their teeth or the curve of their jaws show some similarities with that of man, but those similarities are not unusual according to primordialization theory. Therefore, those skeleton similarities offer no ground to call them the remains of an ape-like human ancestor and as evidence of link between human and apes.

Just to show how speculative the stories about australopithecines are, in this place I consider it appropriate to mention a story about the interpretation of some bone finds ascribed to those so-called human ancestors. At the excavation site of australopithecine at Swartkan from the bone remains of some other

animals piled up beside those of australopithecine one paleon-
tologist concluded that they were resulted from the activities of
bloodthirsty australopithecines, another one looking at the same
bones later found them as leftovers from the meals of carnivores.
In addition, since the bones found in that deposit were black-
ened, this was first thought to be due to the use of fire, later the
deposition of manganese on the bones was found to be the cause
of the blackening.

The only assumption which might possibly be correct about
australopithecines is that such creatures were living 4 to 1 million
years ago on earth, and then for whatever reason they became
extinct. The presence of sharp flints at the sites of australopithecines, even if it might indicate their familiarity with tools, has
no relevance as proof of their relationship with neither man nor
apes.

4. If the human skull and brain were evolved from the skull
and brain of an ape-like creature, it needed to be explained how
it occurred developmentally, and how the interconnection between the development of the two structures were established.
Unfortunately, questions of great significance like this have been
left almost untouched. In the following I will explain that the skull
and brain of man are not evolved from an ape-like creature.

Neither upright walking nor free use of hands can influence the size of the brain; all three are programmed simultaneously

In a study it has been argued that a reduction in sphenoid
bone size may account for a rapid evolution of modern human

cranial shapes. This kind of notion does not recognize the fact that developmentally the size of the skull cannot be determined by random changes in the size of the sphenoid, or any other bone, since factors that determine the size of the head and the brain seem to be developmentally interconnected. Let us ask this hypothetical question: Was it in case of those ape-like creatures their attempts to walk on two limbs, or their attempts to use their hands that led to the enlargement of their brains? None of them, is the answer, because there are no ways by which those attempts could influence the generative cells of those apes to lead to the development of individuals with larger skulls and brains. Studies on vertebrates' embryos indicate that cells from the dorsal ectoderm give rise to neural tissue. As it has been shown in experiments with zebra fish, in order for the dorsal ectoderm cells to develop to brain tissue the cooperation of a small group of cells located in the prospective head region of the gastrula is required. Without these cells the formation of brain tissue is disturbed. This study also indicates that in a developing embryo the factor which determines how large the brain should be depends from the size of the head region. It seems that not only the formation of the brain but also its size is jointly controlled by the cells of the head region. Having this correlation in mind, there is no possibility for any attempt of upright walking, or free use of hands to influence an already established relation of body parts. This cannot be achieved through accumulation of tiny changes. Thus, it is impossible for human to be evolved from apes or ape-like creatures.

Who were Neanderthals?

Neanderthals were living for about 200000 years on earth before they disappeared 40000 years ago (these times are rough estimates). Based on skeleton finds, their skull bones show slight differences in size and shape when compared to ours, but their form is recognizable as a human form, with a brain size of about 1400 cubic centimeters. In comparison, the size of the brain of those ape-like creatures that according to different estimations lived between 4 to 1 million years ago was approximately 400 cubic centimetres. If Neanderthals were the result of a further evolution from australopithecines, separated from each other by about one million years, then how could the 1000 cubic centimetres jump in the size of the Neanderthals' brain be explained? Further, since australopithecines with their 400 cubic centimeters brains were supposedly walking upright and using their hands, then one may conclude that such a rapidly brain enlargement of 1000 cubic centimetres in Neanderthals must have been exclusively related to making some artificial stone tools, something very unthinkable. Moreover, since Neanderthals were humans as we are and had the same brain size as we have, so we should be able to identify this 1000 cubic centimetres tool-making area in our own brain. Where is this area? In fact, such an area in our brain does not exist. One can continue arguing that if Neanderthals for making their tools needed 1000 cubic centimetres larger brains than the australopithecines, then where is the amount of brain enlargement needed for the development of language, agricultural and architectural performances by humans in the last few thousand years. There is, however, no evidence that any of the recent human performances coincides with any brain enlargement. Thus, it must be concluded that no form of activity could make a Neanderthal brain out of an australopithecine brain. Therefore, Neanderthals were emerged through primordialization as humans and, as a matter of fact, had no relation with australopithecines.

From these analyses it becomes clear that the tendencies to interpret phylogenetic relations between fossil finds by putting emphasis on some selectively chosen anatomical features are groundless. If there were any such relations involved in diversification of living organisms, then at least concerning the more recent organisms one should be able to find many examples of intermediate fossils between any two organisms considered to be related. It is also very superficial to put existing living organisms with very different morphological, physiological and behavioural characteristics in one group and call them relatives because of sharing some anatomical or physiological features, such as, for example, the mammary glands in case of mammals; or homoeothermic feature in case of birds and mammals. It is equally not correct to assume that because of the presence of their extra-embryonic membrane (the amnion), reptiles, birds, and mammals are relatives.

SUMMARY OF THE THEORY

1. Emergence of a biological form requires primordialization, which is programming of an urprimordial cell or a cell equivalent to it with the design of the form to be emerged. The urprimordial cell in this process turns into a primordial cell.

2. An urprimordial cell is a cell that contains the basic functions of a living cell, such as growth, reproduction (sexually and asexually), signal transmission, communication with the environment, and the potential for cell-to-cell association.

3. A primordial cell is a cell that in addition to the basic functions and potentials of an urprimordial cell also contains a regulative program that specifies the design of a form, its functions and behavior, and the layout of its structures during the embryonic development. This regulative program is encoded by invariant proteins.

4. An ordinary or an embryonic cell can also be turned into a primordial cell through the process of reprimordialization; whereby after disintegration of its original primordial program the cell acquires a new primordial program, given that none of the invariant proteins that have to take part in the new program is altered. I call the cell in transition equivalent to urprimordial cell.

5. In an ordinary or in an embryonic cell when any of the invariant proteins that participate in its primordial program is altered, the primordial program of that cell destabilizes. A cell

with a destabilized primordial program turns into a cancer cell.

6. Transformation of one form into another is impossible. Form-specific proteins do not exist. Alterations of DNA sequences and variations in size and number of chromosomes are not the cause of diversity of biological forms.

7. Functions in metazoan derive from functions in urprimordial cells.

8. Behavior in living organisms is also programmed during primordialization. Nervous systems in metazoan are the re-structured versions of urprimordial cell's communication and decision-making systems.

9. Cell proteins are intelligent molecules for having the ability to recognize events and memorize their sequence.

10. Adaptability is a potential that allows an organism to undergo functional, behavioral, and morphological changes within the boundary of the form's design.

Conclusion

In this book I presented a comprehensive biological theory that is able to answer a number of still unresolved questions in different areas of biology such as the origin of forms' designs, evolution, biodiversity, development, initiation of cancer, and the origin of animal behavior. Some major aspects of this theory are the following.

1. We want to know what makes the living organisms to be different from each other. What is it at the molecular level that makes, for example, a man to be different from an ape? Why in one instance the same proteins, regulative and non-regulative, produce a form with the shape of a man and in another instance that of an ape, and so on? These are understandably very challenging questions. What are the right answers to them?

Looking at the uniqueness of each biological form, I assumed that in the absence of forms' specific proteins the design of a form must be determined by a regulative system that I call it primordial program. Primordial programs are coded by certain groups of proteins. Each form has its own specific primordial program. In order to fulfill such job, those proteins have to be invariant, undergone no changes in critical parts of their sequences since the primordial times, and can combine with each other in thousands different ways (depending on the number and sequence of their alignment). Primordialization theory postulates that the origin of all essential proteins in protozoan and metazoan are in urprimordial cells. Invariant proteins that I consider very likely candidates for making the primordial programs are among the invariant proteins that determine the polarity of organisms, proteins in charge of segmentation and symmetries in developing embryos, growth factors, growth factors receptors, proteins that are involved in signal transmission, and proteins that are in control of DNA molecules. The mechanism in charge of combining these proteins in specific patterns to produce primordial programs I called primordialization. During this process urprimordials cell are turned into primordial cells, so that different primordial cells have different primordial programs. These programs can also be encoded in cells that are equivalent to urprimordial cells. Interestingly, it has been reported recently that metazoans' proteins such as cell signaling and adhesion proteins can have their origin from coenoflagellates: Expression in these organisms of proteins involved in cell interactions in Metazoa indicates that these proteins already existed before the emergence of animals." When this is the case, then one has to go one step further asking the question from which organisms the coenoflagellates obtained their proteins. Primordialization theory says: from urprimordial cells, certainly not so that those proteins are being passed from protozoan to metazoan, but so that protozoan and metazoan all receive them directly from urprimordial cells or cells equivalent to them at any time when a new distinct organism (form) emerges.

A primordial cell can then develop into an organism. Its primordial program determines the design and the layout of the structures of the organism during its embryonic development. Primordialization is not bound to a certain point in time in the history of living organisms on earth. It is a process by which forms have been designed and emerged on a continuous basis. In the absence of urprimordial cells, other cells that are equivalent to them are being used in this process. Changes of DNA sequences, others than those that encode the invariant proteins of the primordial programs, and variations in size and number of chromosomes have no effect on the design of biological forms. Variable proteins, as the sequence analyses of these molecules from different organisms show, do not seem to be responsible for differences in design of living organisms. In recent years it has been tried by many workers to compare the sequences of proteins from different organisms in order to establish phylogenetic relations between them. Critical evaluation of data, however, indicates that comparisons of proteins sequences from different organisms provide no evidence of a phylogenetic relatedness between them, and none of those variations can account for the basic morphological differences we observe among living organisms. On the contrary, those data can be interpreted in favor of the notion that the design of the biological forms cannot be touched by proteins that are variable. Moreover, there are further findings that are indirectly speak against the notion of phylogenetic relations among living organisms: first, in a number of studies conducted in invertebrates and vertebrates similarities between their subordinate regulative systems that are at work during their development have been shown. Second, with regard to structural proteins there are studies that are indicating that the subordinate regulative systems are using similar structural proteins for building different organs and structures. This means that structural proteins are not made to fit only a certain organism's need. The same structural protein can be used by different organisms. The difference between the formation of a limb of a land vertebrate and the fin of a fish does not require that their structural proteins should be different from each other. Further, in any organism its secondary regulative systems seem to use the same structural proteins for different morphological purposes, depending on the position of those structures in the body. For example, in Drosophila the same structural proteins are used for the formation of antennae and legs, or halters and wings. In the same organism in experiments where the loss of function of some homeotic proteins results in changes of thoracic parts of the body into structures appropriate to abdominal segments, and vice versa, it has been shown that both visceral and thoracic organs can be built from the same structural proteins.

To illustrate the significance of the mentioned findings with regard to the lack of phylogenetic relations among living organisms, that means their having predetermined designs, I tend to compare the subordinate regulative systems in living organisms to a team of workers – masons, carpenters, painters, cement mixers, etc. – who are working together on construction sites with different designs. Those workers' duties are well-defined: They have to perform their jobs within the framework of each construction design. So the job of subordinate regulative systems in each living organism, indeed, is to work along the path of its development according to the instruction of its design regulative system. Concerning the use of similar structural proteins for different structures, I compare those proteins to construction material those workers are using. In a similar way that they can use the same material on different sites or on diverse locations within the same construction site, the subordinate regulative systems do it with the structural proteins.

2. The patterns of proteins combination in primordial programs cannot be shifted one into another by environmental factors. If the amino acid sequence of any of the participating proteins is altered, the design program will obliterate. In case such an alteration happens inside a primordial cell (or egg cell) itself, one of the two possibilities is expected to occur. Either the affected cell is no longer viable and dies out, or develops into a creature with indistinct body design. Alteration of the proteins involved in a primordial program can occur also at any time during an organism's development and later. In such situations the affected cells can turn into a cancer cell. According to this notion, cancer gets a new definition other than the one that is currently in use. In the relevant literature cancer has been considered as a the result of accumulation of mutations and other inheritable changes in susceptible cells, such as loss-of-function mutations (inactivating intragenic mutations, gene deletions, and epigenic silencing) and gain-of-function mutations (activating intragenic mutations, gene amplification, and gene translocations). Using a

different phrase, cancer has been seen as clonal proliferation that arises owing to mutations that confer selective growth advantage on cells. However, we have to keep in mind that 'accumulation of mutations' and 'mutations that confer selective growth advantage on cells' are describing the changes being observed in cancer cells, but do not say anything about how any one of those mutations initiate cancer, i.e. turning a normal cell into a cancer cell. Screening for muted genes in cancer cells lead to finding of more than 350 such genes so far. One may select the most significant ones of those genes and tries step by step to find out the kind of gene or genes that are supposedly offering growth advantage to a cell, and how it woks. The chances are that no mutation with the ability of conferring growth advantage will be found, unless one of the following two situations are given: 1. the existence a cancer making mechanism inside the cells that can be *turned on* by some mutations; this is very unlikely, because it is not in the interest of living organisms to carry such a potential of self-destruction within their cells. 2. Mutations that are supposed to offer growth advantage have to be in position to change the whole chemistry of the affected cell; this is also unthinkable. In my opinion any protein that its mutation prompts a cell to grow out of control is part of its primordial program. Mutations of such proteins by destabilizing the design program of the cell turn it into a defective urprimordial cell. These cells are unable to recognize the boundary of the form, thus grow out of control.

3. A disruption of the pattern of a primordial program by means other than changes in the amino acid sequences of its

constituent proteins can lead to disintegration of a form into undifferentiated cells. These cells, however, can be reprimordialized, that means their invariant proteins can be reprogrammed to design a new form. Some examples of reprimordialization process that have been described in this book are reprimordialization of some embryonic cells of a developing organism, the so-called metamorphosis, and the process such as the one that turns an unicellular *Dictyostelium discoideum* into a multicellular slug.

4. In accordance with the theory described in this book, adaptability is redefined. For the sake of their survival living organisms need to evaluate the unfavorable conditions and circumstances under which they live, unless these are beyond their control such as natural disasters. The essence of this evaluation is to overcome the obstacles and shortages that endanger the means they need to thrive and survive. Since the environment is steadily changing, the threats are changing, too, in form and magnitude. Therefore, living organisms must be able to stand up to those changes. They can do this through changing their behavior, their biochemistry and cell functions. As far as changes of behavior are concerned, they would very likely look around for a safer nesting place, or eventually move to somewhere else, for example. When this deems to be insufficient, biochemical and functional changes have to occur, such as changing of their skin color, camouflage and others. If they fail to do so, they would die. These changes, however, are not dictated by the environment; in other words, biological organisms do not change passively or randomly as it has been generally assumed. They rather change themselves from

within, thanks to the intelligence of their protein molecules. Thus, adaptability is a potential within the organisms which helps them to overcome the effects of environmental changes in order to maintain the integrity of their design.

REFERENCES

Aguinaldo, A. M. A. et al. (1997). Evidence for a clade of nematodes, arthropods and other molting animals. Nature 387, 489-493.

Akam, M. (1989). Hox and HOM: homologous gene clusters in insects and vertebrates. Cell 57, 347-349.

Alberts, B., Bray, D., Lewis, J., Raff, M., Robert, K. and Watson, J. D. (1989). Molecular Biology of the Cell, 2nd ed., Gerland Publishing, New York and London.

Alcock, J. (1989). Animal Behavior, An Evolutionary Approach, 4th ed., Sinauer Associates, Sunderland, Mass.

Anderson, K. V. (1987). Dorsal-ventral embryonic pattern genes of Drosophila. Trends in Genetics. 3, 91-97.

Andersson, D. I., Slechta, E. S. and Roth, J. R. (1998). Evidence that gene amplification underlies adaptive mutability of the bacterial lac operon. Science 282, 1133-1135.

Averof, M. (1998). Origin of the spider's head. Nature 395, 436-437.

Baulcombe, D. C. (1996). RNA as a target and an initiator of post-transcriptional gene silencing in transgenic plants. Plant. Molec. Biol. 32, 79-88 .

Baulcombe, D. (2007). Amplified silencing. Science 315, 197-200.

Benton, M. J. (1990). Phylogeny of the major tetrapod groups: morphological data and divergence dates. Journal of Molecular Evolution 30, 409-424.

Benton, M. J. (2000). Stems, nodes, crown clades, and rand-free list: is Linnaeus dead? Biological Reviews 75, 633-648.

Beukeboom, L. W. et al. (2007). Haploid females in the parasitic wasp Nasonia vitripennis, Science 315, 206.

Blackburn, D. G. (1999). Placenta and placental analogs in reptiles and amphibians. In Encyclopedia of Reproduction (eds. E. Knobil, and J. D. Neil), pp. 840-847. Academic Press, San Diego.

Blackwelder, R. E. and Garoian, G. S. (1986). Handbook of Animal Diversity. CRC Press, Boca Raton, FL.

Bray, D., Levin, M. D. and Morton-Firth, C. J. (1998). Receptor clustering as a cellular mechanism to control sensitivity. Nature 393, 85-88.

Britten, R. J. and Kohne, D. E. (1968). Repeated sequences in DNA. Science 161, 529-540

Brooker, R. J. (1999). Genetic Analysis and Principles. Benjamin Cummings, California.

Camin, J. H. and Sokal, R. R. (1965). A method for deducing branching sequences in phylogeny. Evolution 19, 311-326.

Campos-Ortega, J. A. and Hartenstein, V. (1985). The Embryonic Development of Drosophila melanogaster. Springer, Berlin.

Casares, F. and Man, R. S. (1998). Control of antennal versus leg development in Drosophila. Nature 392, 723-728.

Cheng, T. C. (1964). The Biology of Animal Parasites. Saunders, Philadelphia.

Chirpich, T. P. (1975). Rates of protein evolution: a function of amino acid composition. Science 188, 1022-1023.

Coyne, J. A. and Orr, H. A. (2004). Speciation. Sinauer Associates, Sunderland, Mass.

Darwin, C. R. (1859). On the Origin of Species by Means of Natural Selection. John Murry, London.

DeDuve, C. (1991). Blueprint for a Cell: The Nature and Origin of Life. Neil Patterson, Burlington, NC.

Dell, H. (2007). Marked from the start. Nature 445, 157.

DiBerardino, M. A., Orr, N. H. and McKinnell, R.G. (1986). Feeding tadpoles cloned from Rana erythrocyte nuclei. Proceeding of the National Academy of Sciences, USA 83, 8231-8234.

Dickerson, R. E. (1971). The structure of cytochrome c and the rates of molecular evolution. Journal of Molecular Evolution 1, 26-45.

Dogiel, V. A., Polyanski, Yu. I. and Kheisin, E. M. (1966). General Parasitology. Academic Press, New York.

Driever, W. and Nüsslein-Volhard, C. (1988). The bicoid protein determines position in the Drosophila embryo in a concentration dependent manner. Cell 54, 95-104.

Duboule, D. and Dolle, P. (1989). The structural and functional organization of the maurine Hox gene family resembles that of Drosophila homeotic genes. Journal of European Molecular Biology Organization 8, 1497-1505.

Engel, R. (1990). Mating-type genes, meiosis and sporulation. In: Molecular Biology of the Fission Yeast (eds. A. Nasim, B. Johnson, and P. Young), pp. 32-37. Academic Press, San Diago.

Fitch, W. M. and Margoliash, E. (1967). Construction of phylogenetic trees. Science 155, 279-284.

Frankhauser, G. (1952). Nucleo-cytoplasmic relations in amphibian development. International Review of Cytology 1, 165-193.

Frankhauser, G. (1955). In: Analysis of Development (eds. B. H. Willer, P. A. Weiss, and V. Hamburger), pp. 126-150. Saunders, Philadelphia.

Gehring, W. J. and Nöthiger, R. (1973). The imaginal discs of Drosophila. In: Developmental Systems: Insects (eds. S. Counce and C. H. Waddington), Vol. 2, pp. 211-290, Academic Press, New York.

Gehring, W. J., Müller, M., Affolter, M., Percival-Smith, A., Billeter, M., Qian, Y. Q., Otting, G. and Wüthrich, K. (1990). The structure of the homeodomain and its functional implications. Trends in Genetics. 6, 323-329.

Gibbons, A. (1998). Which of our genes make us humans? Science 281, 1432-1434.

Gibbs, R. A. (2007). Evolutionary and biochemical insights from the rhesus macaque genome. Science 316, 222-234.

Gilbert, S. F. (1994). Developmental Biology, 4th ed., Sinauer Associates, Sunderland, Mass.

Goodman, M., Moore, G. W. and Matsuda, G. (1975). Darwinean evolution in the genealogy of heamoglobin. Nature 253, 603-608.

Graham, A., Papalulu, N. and Krumlauf, R. (1989). The maurine and Drosophila homeobox gene complexes have common features of organization and expression. Cell 57, 367-374.

Greenman, C. et al. (2007). Patterns of somatic mutation in human cancer genomes. Nature 446, 153-158.

Griffith, G. M., Berek, C., Kaartinen, M., and Milstein, C. (1984). Somatic mutation and the maturation of the immune response to 2-phenyl oxazolone. Nature 312, 271-275.

Gurdon, J. B. (1968). Transplanted nuclei and cell differentiation. Scientific American 219 (6), 24-35.

Hammord, S. M., Bernstein, E., Beach, D. and Hannon, G. I. (2000). An RNA-directed nuclease mediates post-transcriptional gene silencing in Drosophila cells. Nature 404, 293-296.

Hannon, G. J. (2002). RNA interferance. Nature 418, 244-251.

Harris, R. A., Rogers, J. and Milosavljevic, A. (2007). Human-specific changes of genomic structures detected by genomic triangulation. Science 316, 235-237.

Hiller, L. W. et al. (2004). Sequence and comparative analyses of the chicken genome provide unique perspective on vertebrate evolution. Nature 432, 695-716.

Hillis, D. M., Moritz, G. and Mable, B. K. (1996). Molecular Systematics. Sinauer Associates, Sunderland, Mass.

Hopkin, M. (2007). Chimps lead evolutionary race. Nature 446, 841.

Hoyer, B. H., McCarthy, B. J. and Bolton, E. T. (1964). A molecular approach in the systematics of higher organisms. Science 144, 959-967.

Ingham, P. W. (1988). The molecular genetics of embryonic pattern formation in Drosophila. Nature 335, 25-34.

Irion, U. and Johnston, D. St. (2007). Bicoid RNA localization requires specific binding of an endosomal sorting complex. Nature 445, 554-558.

Johns, B. and Miklos, G. L. G. (1988). The Eukaryote Genome in

Development and Evolution. Allen and Unwin, London.

Jukes, T. H. and Holmquist, R. (1972). Estimation of evolutionary changes in certain homologous polypeptide chains. Journal of Molecular Biology 64, 163-179.

Kay, R. F., Ross, C. and Williams, B. A. (1997). Anthropoid origins. Science 275, 797-804.

Kazemie, M. (2000). A DNA-independent mechanism of diversity of living organisms. Meeting of the Society for Molecular Biology and Evolution, New Haven, Conn.

Kazemie, M. (2001). Invariant proteins determine the designs of the biological forms. Meeting of the Society for the Study of Evolution, Knoxville, TN.

Kazemie, M. (2006). A new mechanism of cancerogenesis: Destabilization of cell's design regulative system. Cold spring Harbor Meeting on Mechanisms and Models of Cancer. Cold Spring Harbor, NY.

Kimura, M. (1968). Evolutionary rate at the molecular level. Nature 217, 624-626.

King, J. L. and Jukes, T. H. (1969). Non-Darwinian Evolution. Science 164, 788-798.

King, M-C. and Wilson, A. C. (1975). Evolution at two levels in humans and chimpanzees. Science 188, 107-116.

King, N., Hittinger C. T. and Carroll S. B. (2003). Evolution of key cell signalling and adhesion protein families predates animal origin. Science 301, 361-363.

Klobutcher, L. A. and Prescott, D. M. (1986). The special case of the hypotrichs. In: The Molecular Biology of Ciliated Protozoa (ed. J. A. Gall), pp. 111-154. Academic Press, Orlando, FL.

Klug, W. S. and Cummings, M. R. (1997). Concepts of Genetics, 5th ed., Printice Hall, Upper Saddle River, NJ.

Konigsberg, T. R. (1986). The embryonic origin of muscle. In : Myology (eds. A. Engel and B. Q. Banker), Vol. 1, pp. 39-71. McGraw-Hill, New York.

Kuzloff, E. N. (1990). Invertebrates. Saunders College Publishing,

Philadelphia.

Kumar, S. and Hedges, B. (1998). A molecular timescale for vertebrate evolution. Nature 392, 917-920.

Langley, C. H. and Fitch, W. M. (1974). An examination of the constancy of the rate of molecular evolution. Journal of Molecular Evolution 3, 161-177.

Levin, M., Johnson, R. L., Stern, C. D., Kühn, M. and Tabin, C. (1995). A molecular pathway determining left-right asymetry in chick embryogenesis. Cell 82, 803-814.

Loomis, W. F. (1975). *Dictyostelium discoideum.* A Developmental System. Academic Press, New York.

Luo, Zhe-Xi, Li, P. G. and Chen, M. (2007). A new eutriconodont mammal and evolutionary development in early mammals. Nature 446, 288-293.

Mallet, J. (2007). Hybrid speciation. Nature 446, 279-283.

Manning, J. E., Schmidt, C. W. and Davidson, N. (1975). Interspersion of repetitive and nonrepetitive DNA sequences in the *Drosophila melanogaster* genome. Cell 4, 141-155.

Marris, E. (2007). The species and the spacious. Nature 446.

Mayr, E. and Ashlock, P. D. (1991). Principles of Systematic Zoology, 2nd ed., McGraw-Hill, New York.

McKenna, M. C. (1975). Towards a phylogenic classification of the mammalia. In: Phylogeny of the Primates: A Multideciplinary Approach (eds. W. P. Luckett and F .S. Szolay), pp. 21-44. Pleneum Press, New York.

Morris, H., Taylor, G., Masento, M., Jermin, K. and Kay, R. (1987). Chemical structure of the morphogen differntiation inducing factor from *Dictyostelium discoideum.* Nature 328, 811-814.

Müller, W. A. (1997). Developmental Biology. Spinger-Verlag, New York.

Nickoloff, J. A. and Hoekstra, M. F. (1998). DNA Damage and Repair. Humana Press, Totowa, NJ.

Nomura, M. and Li, E. (1998). Smad2 role in mesoderm formation, left-right patterning and craniofacial development. Nature 393, 786-790.

Novacek, M. J. (1992). Mammalian Phylogeny: shaking the tree. Nature 356, 121-125.

Nüsslein-Volhard, C., Frohöfer, H. G. and Lehman, R. (1987). Determination of antero-posterior polarity in Drosophila. Science 238, 1675-1681.

Organ, C. L., Shedlock, A. M., Meade, A., Pagel, M. and Edwards, S. V. (2007). Origin of avian genome size and structure in non-avian dinosaurs. Nature 446, 180-184.

Pennisi, E. (1998). How the genome readies itself for evolution. Science 281, 1131-1134.

Rajewsky, K., Forster, I. and Cumano, A. (1987). Evolutionary and somatic selection of the antibody repertoire in the mouse. Science 238, 1088-1094.

Reddy, S. K., Rape, M., Margansky, W. A. and Kirschner, M. W. (2007). Ubiquitination by the anaphase-promoting complex drives spindle checkpoint inactivation. Nature 446, 921-925.

Rhesus Macaque Genome Sequencing and Analysis Consortium. (2005). Chimp genome sequence. Nature 437, 69-87.

Riggs, A. (1959). Molecular adaptation in haemoglobins: nature of the bohr effect. Nature 183, 1037-1038.

Ryan, A. K., Blumberg, B., Rodriguez-Esteban, C., Yonei-Tamura, S., Tamura, K., Tsukui, T., de la Pena, J., Sabbag, W., Greenwalk, J., Choe, S., Norris, D. P., Robertson, E. J., Evans, R. M., Rosenfeld, M. G. and Izpisua Belmonte, C. (1998). Pitx2 determines left-right asymmetry of internal organs in vertebrates. Nature 394, 545-551.

Schneiderman, H. A. (1976). In: Insect Development (ed. P. A. Lawrence), pp. 3-34, Blackwell, Oxford, UK.

Shephard, J. C. W., McGinnis, W., Carrasco, A. E., De Robertis, E. M. and Gehring, W. J. (1984). Fly and frog homeodomains show homologies with yeast mating-type regulatory proteins. Nature 310, 70-71.

Shinde, U. P., Liu, J. J and Inouye, M. (1997). Protein memory through altered folding mediated by intramolecular chaperones. Nature 389, 520-522.

Simpson, G. G. (1964). Organisms and Molecules in Evolution. Science 146, 1535-1538.

Sjoeblem, T. et al. (2006). The consensus coding of sequences of human breast and colorectal cancers. Science 314, 268-274.

Smith, J. C. (1989). Mesoderm induction and mesoderm-inducing factors in early amphibian development. Development 105, 665-677.

Sokal, S., Wong, G. G. and Melton, D. (1990). A mouse macrophage factor induces head structures and organizes a body axis in Xenopus. Science 249, 561-564.

Stegmeier, F. et al. (2007). Anaphase initiation is regulated by antagonistic ubiquitination and deubiquitination activities. Nature 446, 876-881.

Sternberg, P. W. and Horwitz, H. R. (1986). Pattern formation during vulva development in *C. elegans*. Cell 44, 761-772.

Thompson, J. A, Itskovitz-Eldor, J., Shapiro, S. S., Waknitz, M. A., Swiergiel, J. J., Marshall, V. S. and Jones, J. M. (1998). Embryonic stem cell line derived from human blastocytes. Science 282, 1145-1147.

Torres-Pallida, M-E., Parfitt, D-E., Kouzarides, T. and Zernicka-Goetz, M. (2007). Histone argenine methylation regulates pleuripotency in the early mouse embryo. Nature 445, 214-218.

Vogel, G. (1998). A two-piece protein assembles itself. Science 281, 763-764.

Wang, J. et al. (2007). Opposing LSD1 complexes function in developmental gene activation and repression programs. Nature 446, 882-887.

Webster, G. and Goodwin, B. (1996). Form and Transformation, Generative and Relational Principles in Biology. Cambridge University Press, New York.

Wilkins, A. S. (1993). Genetic Analysis of Animal Development, 2nd ed., Wiley-Liss, New York.

Wellik, D. M. and Capechi, M. R. (2003). Hox 10 and Hox 11 genes are required to globally pattern the mammalian skeleton. Science 301, 363-367.

Whilfield, J. (2007). We are family. Nature 446, 247-249.

Wilson, A. C., Sarich, V. M. and Maxon, L. R. (1974). The importance of gene rearrangements in evolution: evidence from studies on rates of chomosomal, protein, and anatomical evolution. Proceedings of the National Academy of Sciences, USA 71, 3028-3030.

Zuckerkandl, E., Derancourt, J. and Vogel, H. (1971). Multitional trends and random process in the evolution of informational macromolecules. Journal of Molecular Biology 59, 473-490.